Introduction to Engineering Problem Solving

McGraw-Hill's *BEST*—Basic Engineering Series and Tools

Chapman, *Introduction to Fortran 90/95*
D'Orazio and Tan, *C Program Design for Engineers*
Eide et al., *Introduction to Engineering Problem Solving*
Eide et al., *Introduction to Engineering Design*
Eisenberg, *A Beginner's Guide to Technical Communication*
Gottfried, *Spreadsheet Tools for Engineers with Excel*
Mathsoft's, *Student Edition of Mathcad 6.0*
Palm, *Introduction to MATLAB*

INTRODUCTION TO ENGINEERING PROBLEM SOLVING

Arvid R. Eide

Roland D. Jenison

Lane H. Mashaw

Larry L. Northup

Iowa State University

WCB
McGraw-Hill

Boston, Massachusetts   Burr Ridge, Illinois
Dubuque, Iowa   Madison, Wisconsin   New York, New York
San Francisco, California   St. Louis, Missouri

McGraw-Hill

A Division of The **McGraw·Hill** Companies

INTRODUCTION TO ENGINEERING PROBLEM SOLVING

The material in this book is taken from Eide, Jenison, Mashaw, and Northup: *Engineering Fundamentals and Problem Solving.*

This book is printed on acid-free paper.

2 3 4 5 7 8 9 0   DOC/DOC   9 0 9 8

ISBN 0-07-021983-4

Editorial director: *Kevin Kane*
Publisher: *Tom Casson*
Sponsoring editor: *Eric M. Munson*
Developmental editor II: *Holly Stark*
Marketing manager: *John Wannemacher*
Senior project manager: *Jean Lou Hess*
Production supervisor: *Heather D. Burbridge*
Designer: *Larry J. Cope*
Compositor: *York Graphic Services, Inc.*
Typeface: *10/12 Century Schoolbook*
Printer: *R. R. Donnelley & Sons Company*

Library of Congress Cataloging-in-Publication Data

Introduction to engineering problem solving / Arvid R. Eide . . . [et al.].
       p.   cm. — (McGraw-Hill's BEST—basic engineering series and tools)
    Includes bibliographical references and index.
    ISBN 0-07-021983-4
    1. Engineering mathematics.   I. Eide, Arvid R.   II. Series.
TA330.I64   1998
620′.001′51—dc21   97-8464

http://www.mhcollege.com

About the Authors

Arvid R. Eide is a native Iowan. He received his baccalaureate degree in mechanical engineering from Iowa State University in 1962. Upon graduation he spent two years in the U.S. Army as a commissioned officer and then returned to Iowa State as a full-time instructor while completing a master's degree in mechanical engineering. Professor Eide has worked for Western Electric, John Deere, and the Trane Company. He received his Ph.D. in 1974 and was appointed professor and chair of Freshman Engineering at Iowa State from 1974 to 1989. Dr. Eide was selected Associate Dean of Academic Affairs from 1989 to 1995 and currently serves as professor of Mechanical Engineering.

Roland D. Jenison is a professor of aerospace engineering and engineering mechanics at Iowa State University. He has 30 years of teaching experience in lower division engineering and engineering technology. He has taught courses in engineering design graphics and engineering problem solving and has published numerous papers in these areas. His scholarly activities include learning-based instruction in design graphics, design education, and design-build projects for lower division engineering students. He is a long-time member of ASEE and served as chair of the Engineering Design Graphics Division in 1986–1987.

Lane H. Mashaw earned his BSCE from the University of Illinois and MSCE from the University of Iowa. He served as a municipal engineer in Champaign, Ill., Rockford, Ill., and Iowa City, Ia., for nine years and then was in private practice in Decatur, Ill. for another nine years. He taught at the University of Iowa from 1964 to 1974 and at Iowa State University from 1974 until his retirement in 1987. He is currently emeritus professor of civil and construction engineering.

Larry L. Northup is a professor of civil and construction engineering at Iowa State University. He has more than 30 years of teaching experience, with the past 20 years devoted to lower division engineering courses in problem solving, graphics, and design. He has 2 years of industrial experience and is a registered engineer in Iowa. He has been active in ASEE (Engineering Design Graphics Division), having served as chair of the Freshman Year Committee and Director of Technical and Professional Committees (1981–1984). He also served as chair of the Freshman Programs Constituent Committee (now Division) of ASEE in 1983–1984.

Foreword

Engineering educators have had long-standing debates over the content of introductory freshman engineering courses. Some schools emphasize computer-based instruction, some focus on engineering analysis, some concentrate on graphics and visualization, while others emphasize hands-on design. Two things, however, appear certain: no two schools do exactly the same thing, and at most schools, the introductory engineering courses frequently change from one year to the next. In fact, the introductory engineering courses at many schools have become a smorgasbord of different topics, some classical and others closely tied to computer software applications. Given this diversity in content and purpose, the task of providing appropriate text material becomes problematic, since every instructor requires something different.

McGraw-Hill has responded to this challenge by creating a series of modularized textbooks for the topics covered in most first-year introductory engineering courses. Written by authors who are acknowledged authorities in their respective fields, the individual modules vary in length, in accordance with the time typically devoted to each subject. For example, modules on programming languages are written as introductory-level textbooks, providing material for an entire semester of study, whereas modules that cover shorter topics such as ethics and technical writing provide less material, as appropriate for a few weeks of instruction. Individual instructors can easily combine these modules to conform to their particular courses. Most modules include numerous problems and/or projects, and are suitable for use within an active-learning environment.

The goal of this series is to provide the educational community with text material that is timely, affordable, of high quality, and flexible in how it is used. We ask that you assist us in fulfilling this goal by letting us know how well we are serving your needs. We are particularly interested in knowing what, in your opinion, we have done well, and where we can make improvements or offer new modules.

Byron S. Gottfried
Consulting Editor
University of Pittsburgh

Preface

TO THE STUDENT

As you begin the study of engineering you are no doubt filled with enthusiasm, curiosity, and a desire to succeed. Your first year will be spent primarily establishing a solid foundation in mathematics, basic sciences, and communications. You may at times question what the benefits of this background material are and when actual engineering experiences will begin. We believe that they begin now. Additionally, we believe that the material presented in this book will provide you a fundamental understanding of how engineers approach problem solving in today's technological world.

TO THE INSTRUCTOR

Engineering courses for first-year students continue to be in a state of transition. A broad set of course goals, including coverage of prerequisite material, motivation, and retention have spawned a variety of first-year activity. The traditional engineering drawing and descriptive geometry courses have been largely replaced with computer graphics and CAD-based courses. Courses in introductory engineering and problem solving are now utilizing spreadsheets and mathematical solvers in addition to teaching the rudiments of a computer language. The World Wide Web (WWW) is rapidly becoming a major instructional tool, providing a wealth of data to supplement your class notes and textbooks.

Since 1974, students at Iowa State University have taken a computations course that has a major objective of improving problem-solving skills. Various computational aids have been used from programmable calculators to networked PCs. This module has thus evolved from more than 20 years of experience with teaching engineering problem solving to thousands of first-year students.

This module provides procedures for approaching an engineering problem, determining the necessary data and method of solution (including engineering estimates if necessary), and presenting the results using appropriate units. Mathematical expertise beyond algebra, trigonometry, and analytical geometry is not required for any material in the module.

ACKNOWLEDGMENTS

The authors are indebted to many persons who assisted in the development of this module. First we would like to thank the faculty of the former Division of Engineering Fundamentals and Multidisciplinary Design at Iowa State University who taught the engineering computations courses over the past 20 years. They, with support of engineering faculty from other departments, have made the courses a success by their efforts. Several thousands of students have taken the courses, and we want to thank them for their comments and ideas that have influenced this module. The many suggestions of faculty and students alike have provided us with much information necessary to prepare this material. A special thanks to the reviewers whose suggestions were extremely valuable and greatly shaped the manuscript. We also express grateful appreciation to Jane Stowe who worked many hours to type the manuscript. Finally we thank our families for their constant support of our efforts.

Arvid R. Eide
Roland D. Jenison
Lane H. Mashaw
Larry L. Northup

Contents

Engineering Solutions

Introduction

The practice of engineering involves the application of accumulated knowledge and experience to a wide variety of technical situations. Two areas, in particular, that are fundamental to all of engineering are design and problem solving. The professional engineer is expected to intelligently and efficiently approach, analyze, and solve a range of technical problems. These problems can vary from single solution, reasonably simple problems to extremely complex, open-ended problems that require a multidisciplinary team of engineers.

Problem solving is a combination of experience, knowledge, process, and art. Most engineers either through training or experience solve many problem by process. The design process, for example, is a series of logical steps that when followed produce an optimal solution given time and resources as two constraints. The total quality (TQ) method is another example of a process. This concept suggests a series of steps leading to desired results while exceeding customer expectations.

This chapter provides a basic guide to problem analysis, organization, and presentation. Early in your education, you must develop an ability to solve and present simple or complex problems in an orderly, logical, and systematic way.

Problem Analysis

A distinguishing characteristic of a qualified engineer is the ability to solve problems. Mastery of problem solving involves a combination of art and science. By *science* we mean a knowledge of the principles of mathematics, chemistry, physics, mechanics, and other technical subjects that must be learned so that they can be applied correctly. By *art* we mean the proper judgment, experience, common sense, and know-how that must be used to reduce a real-life problem to such a form that science can be applied to its solution. To know when and how rigorously science should be applied and whether the resulting answer reasonably satisfies the original problem is an art.

Much of the science of successful problem solving comes from formal education in school or from continuing education after graduation. But most of the art of problem solving cannot be learned in a formal course; rather, it is a result of experience and common sense. Its application can be more effective, however, if problem solving is approached in a logical and organized method—that is, if it follows a process.

To clarify the distinction, let us suppose that a manufacturing engineer working for an electronics company is given the task of recommending whether the introduction of a new personal computer that will focus on an inexpensive system for the home market can be profitably produced. At the time the engineering task is assigned, the competitive selling price has already been established by the marketing division. Also, the design group has developed working models of the personal computer with specifications of all components, which means that the cost of these components is known. The question of profit thus rests on the cost of assembly. The theory of engineering economy (the science portion of problem solving) is well known by the engineer and is applicable to the cost factors and time frame involved. Once the production methods have been established, the cost of assembly can be computed using standard techniques. Selection of production methods (the art portion of problem solving) depends largely on the experience of the engineer. Knowing what will or will not work in each part of the manufacturing process is a must in the cost estimate, but that data cannot be found in a handbook. It is in the mind of the engineer. It is an art originating from experience, common sense, and good judgment.

Before the solution to any problem is undertaken, whether by a student or by a practicing professional engineer, a number of important ideas must be considered. Consider the following questions: How important is the answer to a given problem? Would a rough, preliminary estimate be satisfactory, or is a high degree of accuracy demanded? How much time do you have and what resources are at your disposal? In an actual situation your answers may depend on the amount of data available or the amount that must be collected, the sophistication of equipment that must be used, the accuracy of the data, the number of people available to assist, and many other factors. Most complex problems require some level of computer support. What about the theory you intend to use? Is it state of the art? Is it valid for this particular application? Do you currently understand the theory, or must time be allocated for review and learning? Can you make assumptions that simplify without sacrificing needed accuracy? Are other assumptions valid and applicable?

The art of problem solving is a skill developed with practice. It is the ability to arrive at a proper balance between the time and resources expended on a problem and the accuracy and validity obtained in the solution. When you can optimize time and resources versus reliability, then problem-solving skills will serve you well.

The engineering method is an example of process. Earlier the engineering design process was mentioned. Although there are different processes that could be listed, a typical process is represented by the following 10 steps:

1. Identification of a need
2. Problem definition
3. Search
4. Constraints
5. Criteria
6. Alternative solutions
7. Analysis
8. Decision
9. Specification
10. Communication

These design steps are simply the overall process that an engineer uses when solving an open-ended problem. One significant portion of this design procedure is step 7—analysis.

Analysis is the use of mathematical and scientific principles to verify the performance of alternative solutions. Analyses conducted by engineers in many design projects normally involve three areas: application of the laws of nature, application of the laws of economics, and application of common sense.

The analysis phase can be used as an example to demonstrate how cyclic the design process is intended to be. Within any given step or phase there are still other processes that can be applied. One very important such process is called the *engineering method*. It consists of six basic steps:

1. Recognize and Understand the Problem

Perhaps the most difficult part of problem solving is developing the ability to recognize and define the problem precisely. This is true at the beginning of the design process and when applying the engineering method to a subpart of the overall problem. Many academic problems that you will be asked to solve have this step completed by the instructor. For example, if your instructor asks you to solve a quadratic algebraic equation but provides you with all the coefficients, the problem has been completely defined before it is given to you and little doubt remains about what the problem is.

If the problem is not well defined, considerable effort must be expended at the beginning in studying the problem, eliminating the things that are unimportant, and focusing on the root problem. Effort at this step pays great dividends by eliminating or reducing false trials, thereby shortening the time taken to complete later steps.

2. Accumulate Data and Verify Accuracy

All pertinent physical facts such as sizes, temperatures, voltages, currents, costs, concentrations, weights, times, and so on must be ascertained. Some problems require that steps 1 and 2 be done simultaneously. In others, step 1 might automatically produce some of the physical facts. Do not mix or confuse these details with data that are suspect or only assumed to be accurate. Deal only with items that can be verified. Sometimes it will pay to actually verify data that you believe to be factual but may actually be in error.

3. Select the Appropriate Theory or Principle

Select appropriate theories or scientific principles that apply to the solution of the problem; understand and identify limitations or constraints that apply to the selected theory.

4. Make Necessary Assumptions

Perfect solutions do not exist to real problems. Simplifications need to be made if they are to be solved. Certain assumptions can be made that do not significantly affect the accuracy of the solution, yet other assumptions may result in a large reduction in accuracy.

Although the selection of a theory or principle is stated in the engineering method as preceding the introduction of simplifying assumptions, there are cases where the order of these two steps should be reversed. For example, if you were solving a material balance problem you often need to assume that the process is steady, uniform, and without chemical reactions, so that the applicable theory can be simplified.

5. Solve the Problem

If steps 3 and 4 have resulted in a mathematical equation (model), it is normally solved by application of mathematical theory, although a trial-and-error solution which employs the use of a computer or perhaps some form of graphical solution may also be applicable. The results will normally be in numerical form with appropriate units.

6. Verify and Check Results

In engineering practice, the work is not finished merely because a solution has been obtained. It must be checked to ensure that it is mathematically correct and that units have been properly specified. Correctness can be verified by reworking the problem using a different technique or by performing the calculations in a different order to be certain that the numbers agree in both trials. The units need to be examined to ensure that all equations are dimensionally correct. And finally, the answer must be examined to see if it makes sense. An experienced engineer will generally have a good idea of the order of magnitude to expect.

If the answer doesn't seem reasonable, there is probably an error in the mathematics, in the assumptions, or perhaps in the theory used. Judgment is critical. For example, suppose that you are asked to compute the monthly payment required to repay a car loan of $5,000 over a 3-year period at an annual interest rate of 12 percent. Upon solving this problem, you arrived at an answer of $11,000 per month. Even if you are inexperienced in engineering economy, you know that this answer is not reasonable, so you should reexamine your theory and computations. Examination and evaluation of the reasonableness of an answer is a habit that you should strive to acquire. Your instructor and employer alike will find it unacceptable to be given results which you have indicated to be correct but are obviously incorrect by a significant percentage.

The engineering method of problem solving as presented in the previous section is an adaptation of the well-known *scientific problem-solving method*. It is a time-tested approach to problem solving that should become an everyday part of the engineer's thought process. Engineers should follow this logical approach to the solution of any problem while at the same time learning to translate the information accumulated in to a well-documented problem solution.

The following steps parallel the engineering method and provide reasonable documentation of the solution. If these steps are properly executed during the solution of problems in this text and all other courses, it is our belief that you will gradually develop an ability to solve a wide range of complex problems.

1. Problem Statement

State as concisely as possible the problem to be solved. The statement should be a summary of the given information, but it must contain all essential material. Clearly state what is to be determined. For example, find the temperature (K) and pressure (Pa) at the nozzle exit.

2. Diagram

Prepare a diagram (sketch or computer output) with all pertinent dimensions, flow rates, currents, voltages, weights, and so on. A diagram is a very efficient method of showing given and needed information. It also is an appropriate way of illustrating the physical setup, which may be difficult to describe adequately in words. Data that cannot be placed in a diagram should be listed separately.

3. Theory

The theory used should be presented. In some cases, a properly referenced equation with completely defined variables is sufficient. At other times, an extensive theoretical derivation may be necessary because the appropriate theory has to be derived, developed, or modified.

4. Assumptions

Explicitly list in complete detail any and all pertinent assumptions that have been made to realize your solution to the problem. This step is vitally important for the reader's understanding of the solution and its limitations. Steps 3 and 4 might be reversed in some problems.

5. Solution Steps

Show completely all steps taken in obtaining the solution. This is particularly important in an academic situation because your reader, the instructor, must have the means of judging your understanding of the solution technique. Steps completed but not shown make it difficult for evaluation of your work and, therefore, difficult to provide constructive guidance.

6. Identify Results and Verify Accuracy

Clearly identify (double underline) the final answer. Assign proper units. An answer without units (when it should have units) is meaningless. Remember, this final step of the engineering method requires that the answer be examined to determine if it is realistic, so check solution accuracy and, if possible, verify the results.

1.5

Standards of Problem Presentation

Once the problem has been solved and checked, it is necessary to present the solution according to some standard. The standard will vary from school to school and industry to industry.

On most occasions your solution will be presented to other individuals who are technically trained, but you should remember that many times these individuals do not have an intimate knowledge of the problem. However, on other occasions you will be presenting technical information to persons with nontechnical backgrounds. This may require methods different from those used to communicate with other engineers, so it is always important to understand who will be reviewing the material so that the information can be clearly presented.

One characteristic of engineers is their ability to present information with great clarity in a neat, careful manner. In short, the information must be communicated accurately to the reader. (Discussion of drawings or simple sketches will not be included in this chapter, although they are important in many presentations.)

Employers insist on carefully prepared presentations that completely document all work involved in solving the problems. Thorough documentation may be important in the event of legal considerations, for which the details of the work might be introduced into the court proceedings as evidence. Lack of such documentation may result in the loss of a case that might otherwise have been won. Moreover, internal company use of the work is easier and more efficient if all aspects of the work have been carefully documented and substantiated by data and theory.

Each industrial company, consulting firm, governmental agency, and university has established standards for presenting technical information. These standards vary slightly, but all fall into a basic pattern, which we will discuss. Each organization expects its employees to follow its standards. Details can be easily modified in a particular situation once you are familiar with the general pattern that exists in all of these standards.

It is not possible to specify a single problem layout or format that will accommodate all types of engineering solutions. Such a wide variety of solutions exists that the technique used must be adapted to fit the information to be communicated. In all cases, however, one must lay out a given problem in such a fashion that it can be easily grasped by the reader. No matter what technique is used, it must be logical and understandable.

We have listed guidelines for problem presentation. Acceptable layouts for problems in engineering are also illustrated. The guidelines are not intended as a precise format that must be followed, but rather as a suggestion that should be considered and incorporated whenever applicable.

Two methods of problem presentation are typical in the academic and industrial environments. Presentation formats can be either freehand or computer generated. As hardware technology and software developments continue to provide better tools, the use of the computer as a method of problem presentation will continue to increase.

If a formal report, proposal, or presentation is to be the choice of communication, a computer-generated presentation is the correct approach. The example solutions that are illustrated in Figs. 1.1, 1.2, and 1.3 include both freehand as well as computer output. Check with your instructor to determine which method is appropriate for your assignments.

The following 9 general guidelines should be helpful as you develop the freehand skills needed to provide clear and complete problem documentation. The first two examples, Figs. 1.1 and 1.2, are freehand illustrations, and the third example, Fig. 1.3, is computer generated. These guidelines are most applicable to freehand solutions, but many of the ideas and principles apply to computer generation as well.

1. One common type of paper frequently used is called engineering-problems paper. It is ruled horizontally, and vertically on the reverse side, with only heading and margin rulings on the front. The rulings on the reverse side, which are faintly visible through the paper, help one maintain horizontal lines of lettering and provide guides for

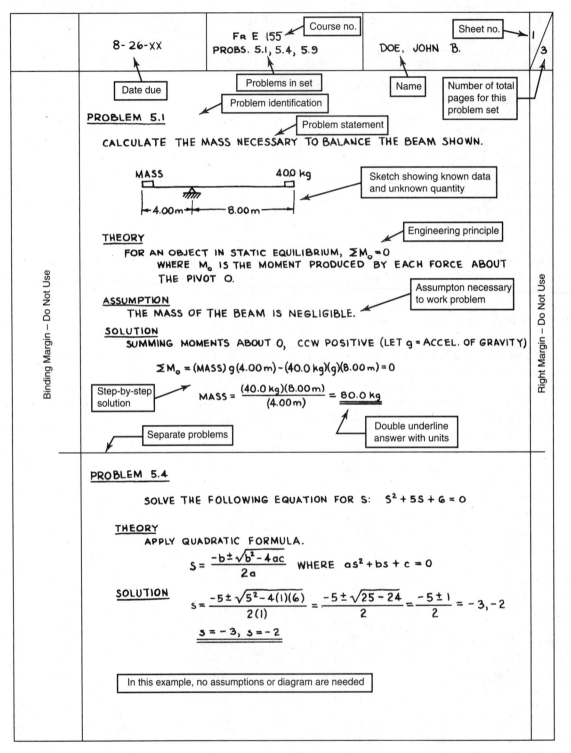

Figure 1.1
Elements of a problem layout.

8

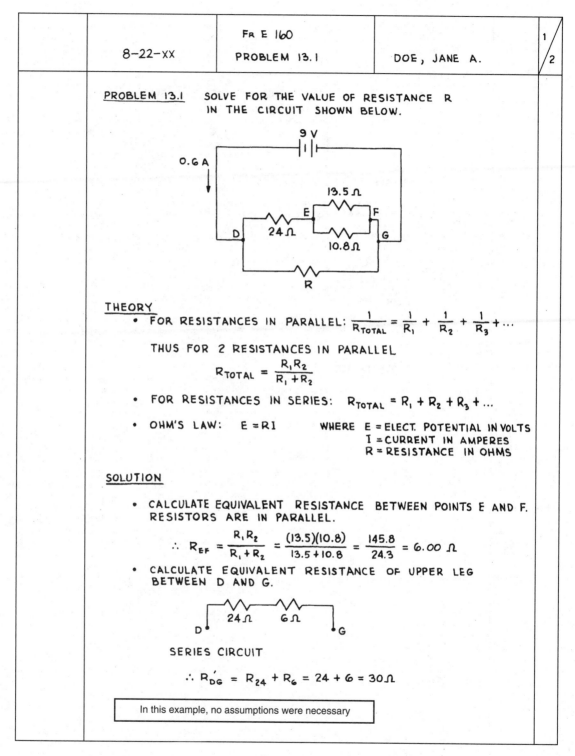

PROBLEM 13.1   SOLVE FOR THE VALUE OF RESISTANCE R
IN THE CIRCUIT SHOWN BELOW.

THEORY
- FOR RESISTANCES IN PARALLEL: $\dfrac{1}{R_{TOTAL}} = \dfrac{1}{R_1} + \dfrac{1}{R_2} + \dfrac{1}{R_3} + \ldots$

  THUS FOR 2 RESISTANCES IN PARALLEL
  $$R_{TOTAL} = \frac{R_1 R_2}{R_1 + R_2}$$

- FOR RESISTANCES IN SERIES: $R_{TOTAL} = R_1 + R_2 + R_3 + \ldots$

- OHM'S LAW: $E = RI$ WHERE E = ELECT. POTENTIAL IN VOLTS
  I = CURRENT IN AMPERES
  R = RESISTANCE IN OHMS

SOLUTION

- CALCULATE EQUIVALENT RESISTANCE BETWEEN POINTS E AND F.
  RESISTORS ARE IN PARALLEL.

  $\therefore R_{EF} = \dfrac{R_1 R_2}{R_1 + R_2} = \dfrac{(13.5)(10.8)}{13.5 + 10.8} = \dfrac{145.8}{24.3} = 6.00 \ \Omega$

- CALCULATE EQUIVALENT RESISTANCE OF UPPER LEG
  BETWEEN D AND G.

  SERIES CIRCUIT

  $\therefore R'_{DG} = R_{24} + R_6 = 24 + 6 = 30 \ \Omega$

  In this example, no assumptions were necessary

Figure 1.2
Sample problem presentation.

9

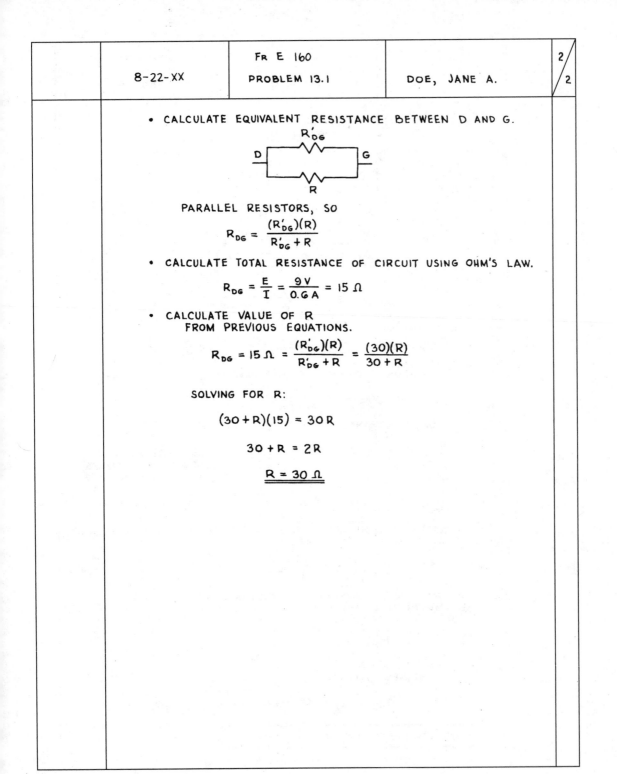

- CALCULATE EQUIVALENT RESISTANCE BETWEEN D AND G.

PARALLEL RESISTORS, SO

$$R_{DG} = \frac{(R'_{DG})(R)}{R'_{DG} + R}$$

- CALCULATE TOTAL RESISTANCE OF CIRCUIT USING OHM'S LAW.

$$R_{DG} = \frac{E}{I} = \frac{9V}{0.6A} = 15 \,\Omega$$

- CALCULATE VALUE OF R
  FROM PREVIOUS EQUATIONS.

$$R_{DG} = 15 \,\Omega = \frac{(R'_{DG})(R)}{R'_{DG} + R} = \frac{(30)(R)}{30 + R}$$

SOLVING FOR R:

$$(30 + R)(15) = 30R$$

$$30 + R = 2R$$

$$\underline{R = 30 \,\Omega}$$

Figure 1.2 (cont.)

Date	Engineering	Name:_____

Problem

A tank is to be constructed that will hold $5.00 \times 10^5 L$ when filled. The shape is to be cylindrical, with a hemispherical top. Costs to construct the cylindrical portion will be $300/$m^2$, while costs for the hemispherical portion are slightly higher at $400 /$m^2$.

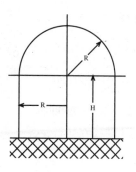

Find

Calculate the tank dimensions that will result in the lowest dollar cost.

Theory

Volume of cylinder is... $V_c = \pi R^2 H$

Volume of hemisphere is... $V_H = \dfrac{2\pi R^3}{3}$

Surface area of cylinder is... $SA_c = 2\pi RH$

Surface area of hemisphere is... $SA_H = 2\pi R^2$

Assumptions

Tank contains no dead air space
Construction costs are independent of size
Concrete slab with hermetic seal is provided for the base.
Cost of the base does not change appreciably with tank dimensions.

Solution

1. Express total volume in meters as a function of height and radius

$$V_{Tank} = f(H, R)$$
$$= V_C + V_H$$
$$500 = \pi R^2 H + \frac{2\pi R^3}{3}$$

Note: $1m^3 = 1000L$

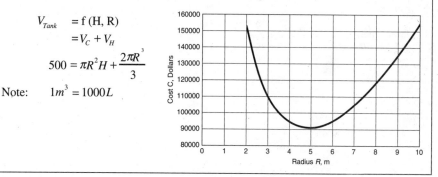

Figure 1.3
Sample problem presentation.

11

2. Express cost in dollars as a function of height and radius

$$C = C\,(H,\,R)$$

$$= 300\,(SA_C) + 400\,(SA_H)$$

$$= 300\,(2\pi RH) + 400\,(2\pi R^2)$$

Note: Cost figures are exact numbers

3. From part #1 solve for $H = H\,(R)$

$$H = \frac{500}{\pi R^2} - \frac{2R}{3}$$

4. Solve cost equation, substituting $H = H\,(R)$

$$C = 300\left[2\pi R\left(\frac{500}{HR^2} - \frac{2R}{3}\right)\right] + 400\left(2\pi R^2\right)$$

$$C = \frac{300000}{R} + 400\pi R^2$$

Cost vs. Radius

Radius R, m	Cost C, $
1.0	301 257
2.0	155 027
3.0	111 310
4.0	95 106
5.0	91 416
6.0	95 239
7.0	104 432
8.0	117 925
9.0	135 121
10.0	155 664

5. Develop a table of Cost vs. Radius and plot graph.

6. From graph select minimum cost.

$$R = \underline{5.00m}$$
$$C = \$91\ 000$$

7. Calculate H from part 3 above

$$H = \underline{3.033\ m}$$

8. Verification / check of results from the calculus:

$$\frac{dC}{dR} = \frac{d}{dR}\left[\frac{300000}{R} + 400\pi R^2\right]$$

$$= \frac{-300000}{R^2} + 800\pi R = 0$$

$$R^3 = \frac{300000}{800\pi}$$

$$R = \underline{4.92m}$$

Figure 1.3 (cont.)

sketching and simple graph construction. Moreover, the lines on the back of the paper will not be lost as a result of erasures.

2. The completed top heading of the problems paper should include such information as name, date, course number, and sheet number. The upper right-hand block should normally contain a notation such as a/b, where a is the page number of the sheet and b is the total number of sheets in the set.

3. Work should ordinarily be done in pencil using an appropriate lead hardness (F or H) so that the linework is crisp and unsmudged. Erasures should always be complete, with all eraser particles removed. Letters and numbers must be dark enough to ensure legibility when photocopies are needed.

4. Either vertical or slant letters may be selected as long as they are not mixed. Care should be taken to produce good, legible lettering but without such care that little work is accomplished.

5. Spelling should be checked for correctness. There is no reasonable excuse for incorrect spelling in a properly done problem solution.

6. Work must be easy to follow and uncrowded. This practice contributes greatly to readability and ease of interpretation.

7. If several problems are included in a set, they must be distinctly separated, usually by a horizontal line drawn completely across the page between problems. Never begin a second problem on the same page if it cannot be completed there. It is usually better to begin each problem on a fresh sheet, except in cases where two or more problems can be completed on one sheet. It is not necessary to use a horizontal separation line if the next problem in a series begins at the top of a new page.

8. Diagrams that are an essential part of a problem presentation should be clear and understandable. Students should strive for neatness, which is a mark of a professional. Often a good sketch is adequate, but using a straightedge can greatly improve the appearance and accuracy of a diagram. A little effort in preparing a sketch to approximate scale can pay great dividends when it is necessary to judge the reasonableness of an answer, particularly if the answer is a physical dimension that can be seen on the sketch.

9. The proper use of symbols is always important, particularly when the International System (SI) of units is used. It involves a strict set of rules that must be followed so that absolutely no confusion of meaning can result. There are also symbols in common and accepted use for engineering quantities that can be found in most engineering handbooks. These symbols should be used whenever possible. It is important that symbols be consistent throughout a solution and that all are defined for the benefit of the reader and for your own reference.

The physical layout of a problem solution logically follows steps similar to those of the engineering method. You should attempt to present the process by which the problem was solved in addi-

**Engineering
Solutions**

tion to the solution so that any reader can readily understand all aspects of the solution. Figure 1.1 illustrates the placement of the information.

Figures 1.2 and 1.3 are examples of typical engineering-problem solutions. You may find that they are helpful guides as you prepare your problem presentations.

1.6

Key Terms and Concepts

The following terms are basic to the material in Chapter 1. You should be able to define these terms and to be able to interpret them into various applications.

Process	Problem presentation
Analysis	Solution documentation
Engineering method	

Problems

The solution to lengths and angles of oblique triangles can be arrived at by application of fundamental trigonometry. All angles are to be considered precise numbers. Solve the following problems using Fig. 1.4 as a general guide.

Figure 1.4

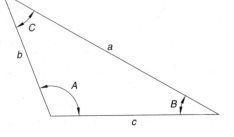

1.1 Given one side and two angles of an oblique triangle.

$C = 75.00°$ $a = 1\,255$ m

$A = 30.00°$

Using laws of sines and sum of angles, determine the angle B and distances b and c.

1.2 Given two sides and the included angle of an oblique triangle.

$C = 40°$ $a = 75$ in

$b = 44$ in

Using the law of cosines and sum of angles, determine the angles A and B and the distance c.

1.3 Given the sides of an oblique triangle.

$a = 440$ ft

$b = 910$ ft

$c = 1\,285$ ft

Using the law of cosines and sum of angles, determine the angles A, B, and C.

1.4 Given two sides and the included angle of an oblique triangle.

$A = 20.00°$ $b = 2\,550$ m

$c = 1\,825$ m

(a) Using the sum of angles, law of sines, and law of tangents, determine the angles B and C and the distance a.
(b) Compute the radius (r), of an inscribed circle and the radius (R) of the circumscribed circle.

Vector quantities have both magnitude and direction. Figure 1.5 illustrates convention for Probs. 1.5 and 1.6, with positive angles measured counterclockwise from the $+ x$-axis.

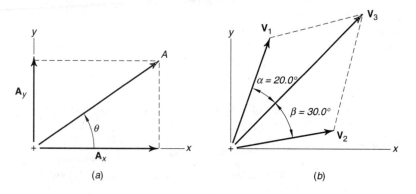

Figure 1.5

(a)   (b)

1.5 A vector **A** has a magnitude of 325 m at $\theta = 52°$ (ccw) from the $+ x$-axis.
   (a) Determine the magnitude of $|Ax|$ and $|Ay|$.
   (b) Determine **A** (magnitude and direction) if $Ax = 85$ m and $Ay = 33$ m.

1.6 A wind vector V_3 is the sum of vectors V_1 and V_2. Vector V_2 makes angle of 15.0° with the $+ x$-axis and has a magnitude of 20.0 mph. Determine the magnitude and direction of vector V_3.

1.7 A survey crew determined the angles and distances given in Fig. 1.6. Calculate the distance across the lake at DE.

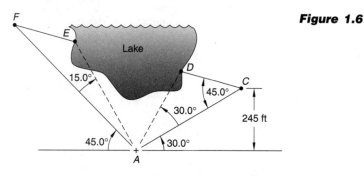

Figure 1.6

1.8 An engineering student in a stationary hot-air balloon is momentarily fixed at 1,325-ft elevation above a level piece of land. The pilot looks down (60° from horizontal) and turns laterally 360°. How many acres are contained within the cone generated by his line of sight? How high would the balloon be if, when performing the same procedure, an area 4 times greater is encompassed?

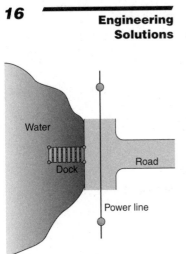

Figure 1.7

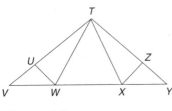

Figure 1.8

1.9 A small aircraft has a glide ratio of 15:1. (This glide ratio means the plane moves 15 units horizontally for each 1 unit in elevation.) You are exactly in the middle of a 3.0 mile diameter lake at 5.00×10^2 ft when the fuel supply is exhausted. You see a gravel road by the dock (Fig. 1.7) and must decide whether to try a ground or water landing. Show appropriate assumptions and calculations to support your decision.

1.10 A pilot in an ultralight knows that her aircraft in landing configuration will glide 2.0×10^1 km from a height of 2.0×10^3 m. A TV transmitting tower is located in a direct line with the local runway. If the pilot glides over the tower with 3.0×10^1 m to spare and touches down on the runway at a point 6.5 km from the base of the tower, how high is the tower?

1.11 A simple roof truss design is shown in Fig. 1.8. The lower section *VWXY* is made from three equal-length segments. *UW* and *XZ* are perpendicular to *VT* and *TY*, respectively. If *VWXY* is 24 m, and the height of the truss is 12 m, determine the lengths of *XT* and *XZ*.

1.12 If you are traveling at 65 mph, what is your angular axle speed in RPM. Assume a tire size of P235/75R15, which has an approximate diameter of 30.0 in.

1.13 The wheel of an automobile turns at the rate of 195 RPM. Express this angular speed in (a) revolutions per second and (b) radians per second. If the wheel has a 30.0-in diameter, what is the velocity of the auto in miles per hour?

1.14 A bicycle wheel has a 28.0-in diameter wheel and is rotating 195 RPM. Express this angular speed in radians per second. How far will the bicycle travel (miles) in 45.0 minutes and what will be the velocity in miles per hour?

1.15 Assume the earth's orbit to be circular at a radius of 93.0×10^6 mi about the sun. Determine the speed of the earth (in miles per second) around the sun if there are exactly 365 days per year.

1.16 A child swinging on a tree rope 8.00 m long reaches a point 4.00 m above the lowest point. Through what total arc in degrees has the child passed? What distance in meters has the child swung?

1.17 Two engineering students were assigned the job of measuring the height of an inaccessible cliff (Fig. 1.9). The angles and distances shown were measured on a level beach in a vertical plane due south of the cliff. Determine the horizontal and vertical distances from *A* to *B*.

Figure 1.9

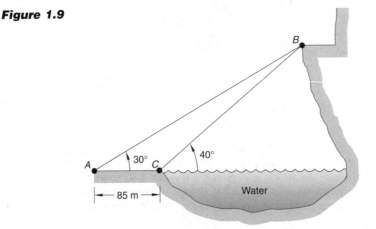

1.18 A survey team with appropriate equipment has been asked to measure a nonrectangular plot of land $ABCD$ (Fig. 1.10). The following data were recorded: $CD = 165.0$ m, $DA = 180.0$ m, and $AB = 115.0$ m. Angle $DAB = 120.0°$ and angle $DCB = 100.0°$.

    (a) Calculate the length of side BC and the area of the plot.

    (b) Estimate the water surface area within the plot and list assumptions.

    (c) Using answers from part b, what percentage of the total surface area within $ABCD$ is land?

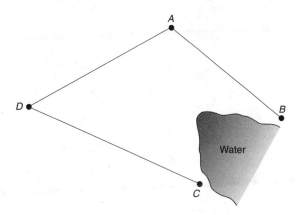

Figure 1.10

1.19 In Fig. 1.11 points S, T, U, and V are survey markers in a land development located on level terrain. The distance ST measures 355.0 ft. Angles at each marker were recorded as follows: $STU = 80.00°$, $TUS = 70.00°$, $VUS = 50.00°$, and $UVS = 70.00°$.

    (a) Calculate the distances TU, UV, VS, and SU.

    (b) Determine the area of $STUV$ in acres.

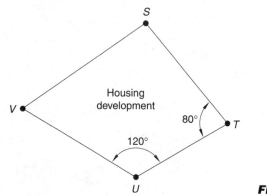

Figure 1.11

1.20 Circular sheets of metal 70.0 cm in diameter are to be used for stamping new highway signs. Calculate the percent of waste when the largest possible inscribed shape is:

    (a) An equilateral triangle

    (b) A square

    (c) An octagon

1.21 Three circles are tangent to each other as in Fig. 1.12. The respective radii are 1.65×10^3, 10.00×10^2, and 7.75×10^2 mm.

    (a) Find the area of the triangle (in square millimeters) formed by joining the 3 centers.

    (b) Determine the area within the triangle that is outside of the circles.

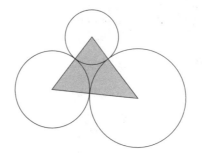

Figure 1.12

1.22 A waterwheel turns a belt on a drive wheel for a flour mill. The pulley on the waterwheel is 2.00 m in diameter, and the drive pulley is 0.500 m in diameter. If the centers of the pulleys are 4.00 m apart, calculate the length of belt needed.

1.23 A narrow, flat belt is used to drive a 50.00 cm (diameter) pulley from a 10.00 cm (diameter) pulley. The centers of the 2 pulleys are 50.00 cm apart. How long must the belt be if the pulleys rotate in the same direction? In opposite directions?

1.24 A homeowner decides to install a family swimming pool. It is 32.0 ft in diameter with a 4.00 ft water level. The cross-section of the pool is illustrated in Fig. 1.13. Consider the following problems.

    (a) How many cubic feet of soil must be excavated to accommodate the pool liner if the bottom profile is as illustrated in Fig. 1.13?

    (b) How many gallons of water will be required to fill the pool to 4 ft above ground?

    (c) If the owner moves the water from a nearby lake, how many tons will be carried? (Density of water is 62.4 lbm/ft³.)

    (d) If the owner carries two 5-gal pails per trip and makes 20 trips per evening, how many days will it take to fill the pool?

Figure 1.13

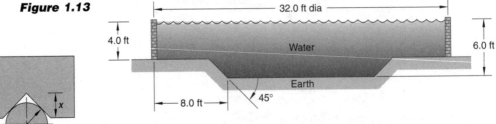

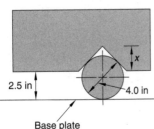

2.5 in

4.0 in

Base plate

Figure 1.14

1.25 A block of metal has a 90° notch cut from its lower surface. The notched part rests on a circular cylinder 4.0 in in diameter as shown in Fig. 1.14. If the lower surface of the block is 2.5 in above the base plate, how deep is the notch?

1.26 Show that the area of the shaded segment in Fig. 1.15 is given by the expression

$$As = \frac{r^2}{2}(\phi - \sin \phi)$$

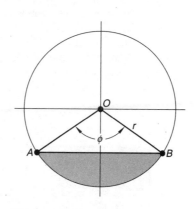

Figure 1.15

1.27 A fuel tank is 10.00 ft in diameter and 20.00 ft long. The tank is buried vertically as shown in Fig. 1.16. Develop an expression for volume (V) in gallons for any depth (h) in feet for this configuration.

1.28 Consider the fuel tank in Prob. 2.27 positioned on its side instead of its base. The tank length is 30.0 ft.

(a) Develop an expression for volume (V) in gallons as a function of radius (r) in feet and height (h) in feet.

(b) If the empty tank with radius 5.00 ft has a mass of 15.00×10^2 lbm, develop an expression for the mass of tank and fuel (in pounds-mass) as a function of height (h) in feet. Density of the fuel is 815 kg/m³.

(c) How many gallons of fuel are in the tank at a height (h) of 4.50 ft? At that depth, what is the total mass of tank and fuel?

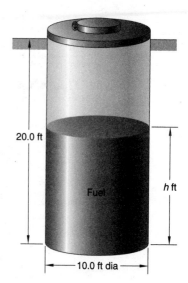

Figure 1.16

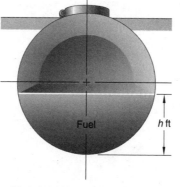

*End view of buried cylindrical tank.

Figure 1.17

1.29 Eighteen circular wooden bases are to be cut from a piece of 3/4-in plywood. Each circular base has a diameter of 10.00 in. Assume a negligible saw blade thickness for the following problems.

(*a*) What is the area of triangle *ABC*?

(*b*) Calculate the area of waste material between *X* and *Y* above *BC*.

(*c*) Determine the dimensions of the rectangular piece of plywood shown in Fig. 1.18 to the nearest 0.5 in.

(*d*) What is the area of the largest piece of waste material?

(*e*) What is the percentage of waste given the configuration shown in Fig. 1.18?

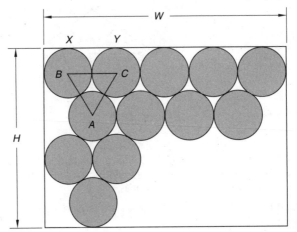

Figure 1.18

Representation of Technical Information

This chapter contains examples and guidelines as well as helpful information that will be needed when collecting, recording, plotting, and interpreting technical data. Two areas will be considered in considerable detail: (1) graphical presentation of scientific data and (2) graphical analysis of plotted data.

Graphical presentation of technical data is necessary when calculated or experimental results are recorded in tabular form. Rapid and accurate determination of relationships between numerical values when the information is reported in columns and rows is not an optimal method for understanding. A procedure for graphing the results is needed. This approach provides a visual impression that is a more intuitive method to compare variables, rates of change, or relative magnitudes.

However, complete graphical analysis involves correct and accurate interpretation of data after it has been plotted. At times impressions are not sufficient, so determination of a mathematical model is required.

Computers with their expanding power, versatility, and speed are changing the way we collect, record, display, and analyze data. For example, consider the rapidly rising popularity of the microprocessor. It has changed the way we do many tasks. Microprocessors or small CPU cards can be programmed by a remote host computer, normally a PC. These CPUs are interfaced with some device, perhaps a temperature sensor or a robot. Information can be collected and stored and then based on the feedback of the information to the CPU additional programmed instructions can direct the device to perform corrective actions.

In addition to microprocessors and various other types of data collection instrumentation, there exists a wide variety of commercial software that enables the engineer to reduce the time required for recording, plotting, and analyzing data while increasing the accuracy of the results. Numerous software programs

Table 2.1

Height H, m	Temp T°C	Pressure P, kPa
0	15.0	101.3
300	12.8	97.7
600	11.1	94.2
900	8.9	90.8
1200	6.7	87.5
1500	5.0	84.3
1800	2.8	81.2
2100	1.1	78.2
2400	−1.1	75.3
2700	−2.8	72.4
3000	−5.0	68.7
3300	−7.2	66.9
3600	−8.9	64.4
3900	−11.1	61.9

are available for both technical presentation as well as analysis.
These programs provide a wide range of powerful tools.

2.1.1
Software for Recording and Plotting Data

Data are recorded in the field as shown in Tab. 2.1. Many times
a quick, freehand plot of the data is produced to provide a visual
impression of the results while still in the field (see Fig. 2.1).

Upon returning to the laboratory, however, spreadsheet software, such as EXCEL and LOTUS 123 provide enormous record-

Figure 2.1
Freehand plot of data.

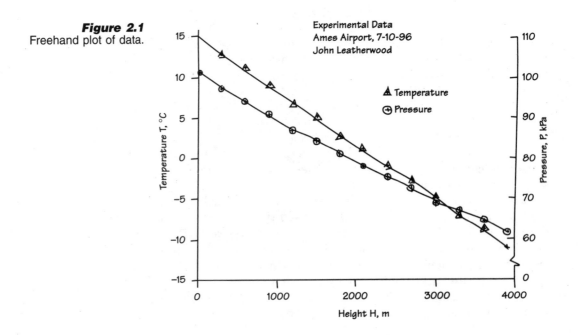

Table 2.2

Height H, m	Temperature $T°C$	Pressure P, kPa
0	15.0	101.3
300	12.8	97.7
600	11.1	94.2
900	8.9	90.8
1 200	6.7	87.5
1 500	5.0	84.3
1 800	2.8	81.2
2 100	1.1	78.2
2 400	1.1	75.3
2 700	−2.8	72.4
3 000	−5.0	68.7
3 300	−7.2	66.9
3 600	−8.9	64.4
3 900	−11.1	61.9

ing and plotting capability. The data are entered into the computer and by manipulation of software options both the data and a graph of the data can be configured, stored, and printed (Tab. 2.2 and Fig. 2.2 and Fig. 2.3).

Programs, such as Mathematica, Matlab, Mathcad, TK Solver, and many others, provide a range of powerful tools designed to help analyze numerical and symbolic operations as well as present a visual image of the results.

Software is also widely available to provide methods of curve fitting once the data has been collected and recorded. This subject will be treated in a separate chapter.

Even though it is important for the engineer to interpret, analyze, and communicate different types of data, it is not practical to include in this chapter all forms of graphs and charts that may be encountered. For that reason, popular-appeal or adver-

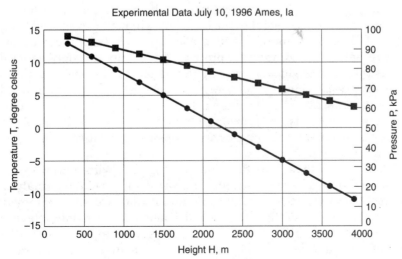

Figure 2.2
Hard copy of computer software analysis.

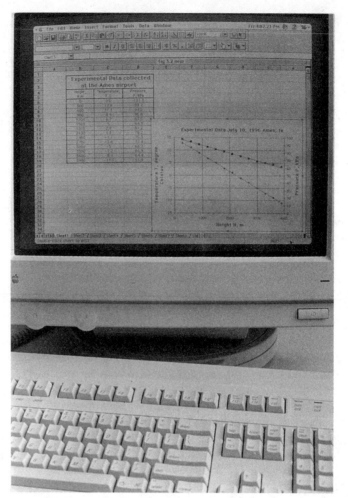

Figure 2.3

tising charts such as bar charts, pie diagrams, and distribution charts, although useful to the engineer, will not be discussed here.

Although commercial software is extremely helpful during the presentation and analysis process, the results are only as good as the original software design and its use by the operator. Some software provides a wide range of tools but only allows limited data applications and minimal flexibility to modify default outputs. Other software provides a high degree of in-depth analysis for a particular subject area with considerable latitude to adjust and modify parameters.

It is inevitable that the computer together with a growing array of software will continue to provide an invaluable analysis tool. However, it is absolutely essential that users be extremely knowledgeable of the software and demonstrate considerable care when manipulating the data. It is important to understand the software limitations and accuracies, but, above all, the operator must know what plotted results are needed and what expectations for appearance and readability are mandated.

It is for this reason that the sections to follow are a combination of manual collection, recording, plotting, and analysis and computer-assisted collection, recording, plotting, and analysis. As we begin the learning process, it is important to know how to mechanically manipulate data so that computer results can be appropriately modified.

2.2
Collecting and Recording Data

2.2.1
Manual

Modern science was founded on scientific measurement. Meticulously designed experiments, carefully analyzed, have produced volumes of scientific data that have been collected, recorded, and documented. For such data to be meaningful, however, certain laboratory procedures must be followed. Formal data sheets, such as those shown in Fig. 2.4, or laboratory notebooks should be used to record all observations. Information about equipment, such as the instruments and experimental apparatus used, should be recorded. Sketches illustrating the physical arrangement of equipment can be very helpful. Under no circumstances should observations be recorded elsewhere or data points erased. The data sheet is the "notebook of original entry." If there is reason for doubting the value of any entry, it may be canceled (that is, not considered) by drawing a line through it. The cancellation should be done in such a manner that the original entry is not obscured, in case you want to refer to it later.

Sometimes a measurement requires minimal precision, so time can be saved by making rough estimates. As a general rule, however, it is advantageous to make all measurements as precisely as time and the economics of the situation will allow. Unfortunately, as different observations are made throughout any experiment, some degree of inconsistency will develop. Errors enter into all experimental work regardless of the amount of care exercised.

It can be seen from what we have just discussed that the analysis of experimental data involves not only measurements and collection of data but also careful documentation and interpretation of results.

Experimental data once collected are normally organized into some tabular form, which is the next step in the process of analysis. Data, such as that shown in Tab. 2.1, should be carefully labeled and neatly lettered so that results are not misinterpreted. This particular collection of data represents atmospheric pressure and temperature measurements recorded at various altitudes by students during a flight in a light aircraft.

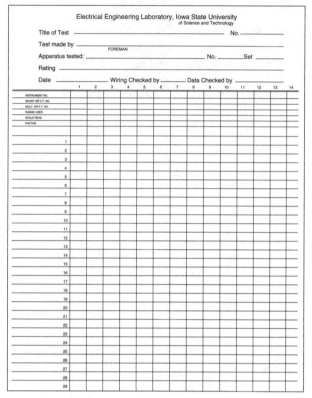

Electrical Engineering Laboratory, Iowa State University
of Science and Technology

Title of Test _____ No. _____

Test made by _____
FOREMAN

Apparatus tested: _____ No. _____ Set _____

Rating _____

Date _____ Wiring Checked by _____ Data Checked by _____

(a)

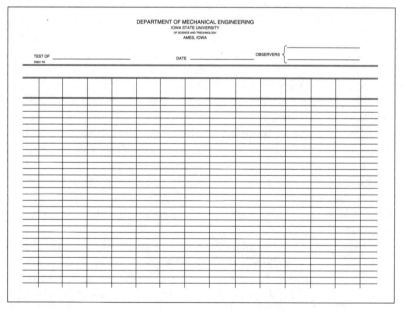

DEPARTMENT OF MECHANICAL ENGINEERING
IOWA STATE UNIVERSITY
OF SCIENCE AND TRECHNOLOGY
AMES, IOWA

TEST OF _____   DATE _____   OBSERVERS {_____

(b)

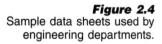

Figure 2.4
Sample data sheets used by
engineering departments.

Although the manual tabulation of data is frequently a necessary step, you will sometimes find it difficult to visualize a relationship between variables when simply viewing a column of numbers. A most important step in the sequence from collection to analysis is, therefore, the construction of appropriate graphs or charts.

2.2.2
Computer Assisted

In recent years, a variety of equipment has been developed which will automatically sample experimental data for analysis. We expect to see expansion of these techniques along with continuous visual displays that will allow us to interactively control the experiments. As an example the flight data collected onboard the aircraft can be entered directly into a spreadsheet and printed as in Tab. 2.2.

Many examples will be used throughout this chapter to illustrate methods of graphical presentation because their effectiveness depends to a large extent on the details of construction.

The proper construction of a graph from tabulated data can be generalized into a series of steps. Each of these steps will be discussed and illustrated in considerable detail in the following subsections.

1. Select the correct type of graph paper and grid spacing.

2. Choose the proper location of the horizontal and vertical axes.

3. Determine the scale units for each axis so that the data can be appropriately displayed.

4. Graduate and calibrate the axes.

5. Identify each axis completely.

6. Plot points and use permissible symbols (that is, ones commonly used and easily understood).

7. Draw the curve or curves.

8. Identify each curve, add title and the other necessary notes.

9. Darken lines for good reproduction.

2.3.1
Graph Paper

Printed coordinate graph paper is commercially available in various sizes with a variety of grid spacing. Rectilinear ruling can be purchased in a range of lines per inch or lines per centimeter, with an overall paper size of 8.5 × 11 inches considered most typical. Figure 2.5a is an illustration of graph paper having 10 lines per centimeter.

Closely spaced coordinate ruling is generally avoided for results that are to be printed or photoreduced. However, for accu-

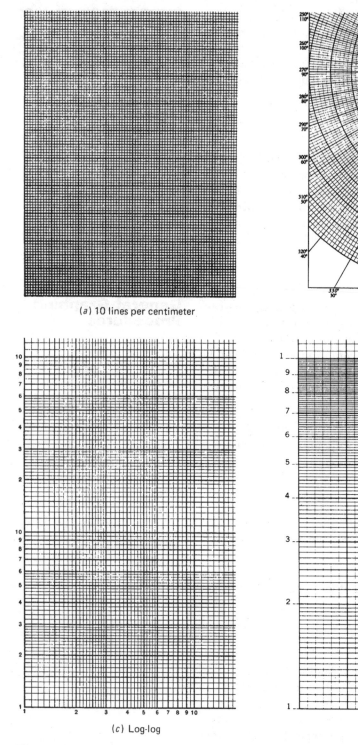

(a) 10 lines per centimeter

(b) Polar graph

(c) Log-log

(d) Semilog

Figure 2.5
Commercial graph paper.

rate engineering analyses requiring some amount of interpolation, data are normally plotted on closely spaced, printed coordinate paper. Graph paper is available in a variety of colors, weights, and grades. Translucent paper can be used when the reproduction system requires a material that is not opaque.

If the data require the use of log-log or semilog paper, such paper can also be purchased in different formats, styles, weights, and grades. Both log-log and semilog grids are available in from 1 to 5 cycles per axis. (A later section will discuss different applications of log-log and semilog paper.) Polar-coordinate paper is available in various sizes and graduations. A typical sheet is shown in Fig. 2.5b. Examples of commercially available logarithmic paper are given in Figs. 2.5c and d.

2.3.2
Axes Location and Breaks

The axes of a graph consist of two intersecting straight lines. The horizontal axis, normally called the *x-axis,* is the *abscissa.* The vertical axis, denoted by the *y-axis,* is the *ordinate.* Common practice is to place the independent values along the abscissa and the dependent values along the ordinate, as illustrated in Fig. 2.6.

Many times, mathematical graphs contain both positive and negative values of the variables. This necessitates the division of the coordinate field into 4 quadrants, as shown in Fig. 2.7. Positive values increase toward the right and upward from the origin.

On any graph, a full range of values is desirable, normally beginning at zero and extending slightly beyond the largest value. To avoid crowding, the entire coordinate area should be used as completely as possible. However, certain circumstances require special consideration to avoid wasted space. For example, if val-

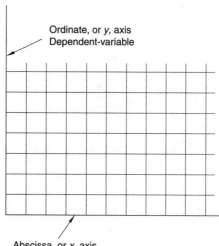

Ordinate, or *y*, axis
Dependent-variable

Abscissa, or *x*, axis
Independent-variable

Figure 2.6
Abscissa(*x*) and ordinate(*y*) axes.

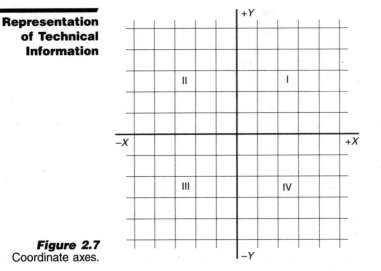

Figure 2.7
Coordinate axes.

ues to be plotted along the axis do not range near zero, a "break" in the grid or the axis may be used, as shown in Figs. 2.8a and b.

When judgments concerning relative amounts of change in a variable are required, the axis or grid should not be broken or the zero line omitted, with the exception of time in years, such as 1970, 1971, and so on, since that designation normally has little relation to zero.

Since most commercially prepared grids do not include sufficient border space for proper labeling, the axes should preferably be placed 20 to 25 mm (approximately 1 in) inside the edge of the printed grid in order to allow ample room for graduations, calibrations, axes labels, reproduction, and binding. The edge of the grid may have to be used on log-log paper, since it is not always feasible to move the axis. However, with careful planning, the vertical and horizontal axes can be repositioned in most cases, depending on the range of the variables.

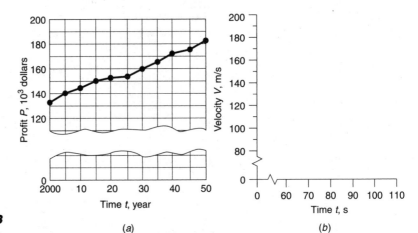

Figure 2.8
Typical axes breaks.

(a)

(b)

2.3.3
Scale Graduations, Calibrations, and Designations

The scale is a series of marks, called *graduations*, laid down at predetermined distances along the axis. Numerical values assigned to significant graduations are called *calibrations*.

A scale can be uniform, with equal spacing along the stem, as found on the metric, or engineer's, scales. If the scale represents a variable whose exponent is not equal to 1 or a variable that contains trigonometric or logarithmic functions, the scale is called a *nonuniform,* or *functional scale.* Examples of both these scales together with graduations and calibrations are shown in Fig. 2.9. When plotting data, one of the most important considerations is the proper selection of scale graduations. A basic guide to follow is the *1, 2, 5 rule,* which can be stated as follows:

Scale graduations are to be selected so that the smallest division of the axis is a positive or negative integer power of 10 times 1, 2, or 5.

The justification and logic for this rule are clear. Graduation of an axis by this procedure makes possible interpolation of data between graduations when plotting or reading a graph. Figure 2.10 illustrates both acceptable and nonacceptable examples of scale graduations.

Violations of the 1, 2, 5 rule that are acceptable involve certain units of time as a variable. Days, months, and years can be graduated and calibrated as illustrated in Fig. 2.11.

Figure 2.9
Scale graduations and calibrations.

Figure 2.10
Acceptable and nonacceptable scale graduations.

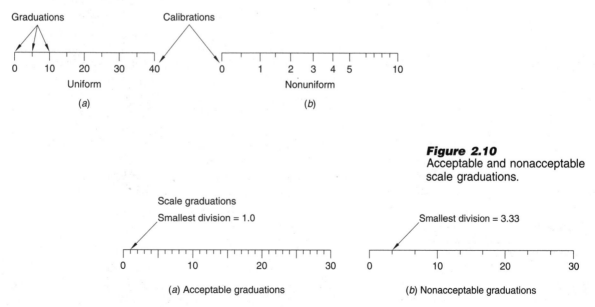

(a) Acceptable graduations   *(b) Nonacceptable graduations*

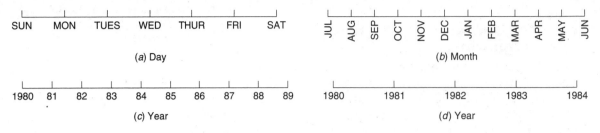

Figure 2.11
Time as a variable.

Scale graduations normally follow a definite rule, but the number of calibrations to be included is primarily a matter of good judgment. Each application requires consideration based on the scale length and range as well as the eventual use. Figure 2.12 demonstrates how calibrations can differ on a scale with the same length and range. Both examples obey the 1, 2, 5 rule, but as you can see, too many closely spaced calibrations make the axis difficult to read.

The selection of a scale deserves attention from another point of view. If the rate of change is to be depicted accurately, then the slope of the curve should represent a true picture of the data. By contracting or expanding the axis or axes, an incorrect impression of the data could be implied. Such a procedure is to be avoided. Figure 2.13 demonstrates how the equation $Y = X$ can be misleading if not properly plotted. Occasionally distortion is desirable, but it should always be carefully labeled and explained to avoid misleading conclusions.

If plotted data consist of very large or small numbers, the SI prefix names (milli-, kilo-, mega-, and so on) may be used to simplify calibrations. As a guide, if the numbers to be plotted and calibrated consist of more than three digits, it is customary to use the appropriate prefix; an example is illustrated in Fig. 2.14.

The length scale calibrations in Fig. 2.14 contain only two digits, but the scale can be read by understanding that the distance between the first and second graduation (0 to 1) is a kilometer; therefore, the calibration at 10 represents 10 km.

Certain quantities, such as temperature in degrees Celsius and altitude in meters, have traditionally been tabulated without the use of prefix multipliers. Figure 2.15 depicts a procedure by which these quantities can be conveniently calibrated. Note in particular that the distance between 0 and 1 on the scale represents 1 000°C. This is another example of how the SI notation is convenient, since the prefix multipliers (micro-, milli-, kilo-, mega-, and so on) allow the calibrations to stay within the three-digit guideline.

Figure 2.12
Acceptable and nonacceptable scale calibrations.

0   10   20

(a) Easy to read

0 1 2 3 4 5 6 7 8 9 10 11 12 13 14 15 16 17 18 19 20

(b) Too crowded

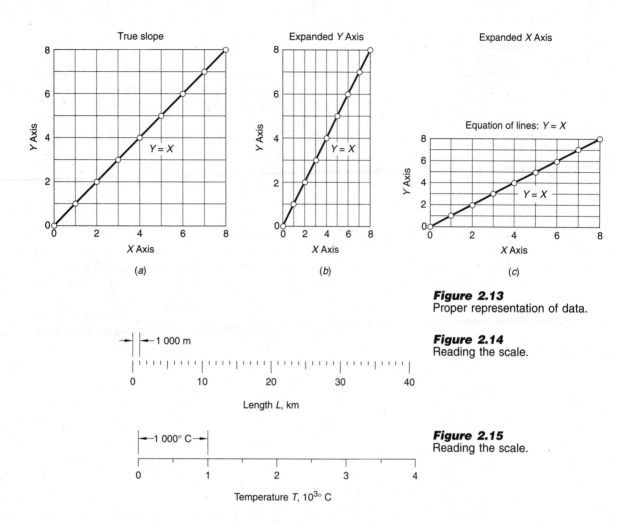

Figure 2.13
Proper representation of data.

Figure 2.14
Reading the scale.

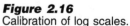

Figure 2.15
Reading the scale.

The calibration of logarithmic scales is illustrated in Fig. 2.16. Since log-cycle designations start and end with powers of 10 (that is, 10^{-1}, 10^0, 10^1, 10^2, and so on) and since commercially purchased paper is normally available with each cycle printed 1 through 10, Figs. 2.16a and b demonstrate two preferred methods of calibration.

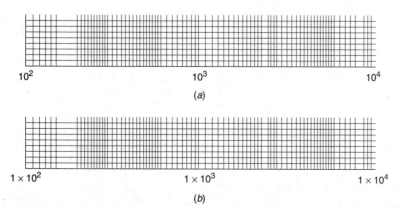

Figure 2.16
Calibration of log scales.

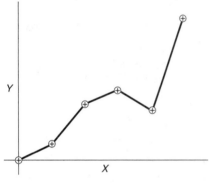

Figure 2.17
Axis identification.

2.3.4
Axis Labeling

Each axis should be clearly identified. At a minimum, the axis label should contain the name of the variable, its symbol, and its units. Since time is frequently the independent variable and is plotted on the x axis, it has been selected as an illustration in Fig. 2.17. Scale designations should preferably be placed outside the axes, where they can be shown clearly. Labels should be lettered parallel to the axis and positioned so that they can be read from the bottom or right side of the page as illustrated in Fig. 2.21.

2.3.5
Point-Plotting Procedure

Data can normally be categorized in one of three general ways: as observed, empirical, or theoretical. Observed and empirical data points are usually located by various symbols, such as a small circle or square around each data point, whereas graphs of theoretical relations (equations) are normally constructed smooth, without use of symbol designation. Figure 2.18 illustrates each type.

Figure 2.18
Plotting data points.

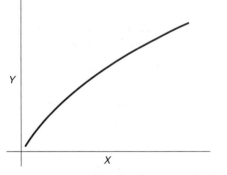

(*a*) Observed: Usually plotted with observed data points connected by straight, irregular line segments. Line does not penetrate the circles.

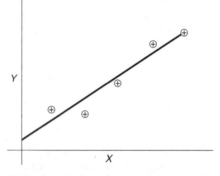

(*b*) Empirical: Reflects the author's interpretation of what occurs between known data points. Normally represented as a smooth curve or straight line fitted to data. Data points may or may not fall on curve.

(*c*) Theoretical: Graph of an equation. Curves or lines are smooth and without symbols. Every point on the curve is a data point.

2.3.6
Curves and Symbols

On graphs prepared from observed data resulting from laboratory experiments, points are usually designated by various symbols (see Fig. 2.19). If more than one curve is plotted on the same grid, several of these symbols may be used (one type for each curve). To avoid confusion, however, it is good practice to label each curve. When several curves are plotted on the same grid, another way they can be distinguished from each other is by using different types of lines, as illustrated in Fig. 2.20. Solid lines are normally reserved for single curves, and dashed lines are commonly used for extensions; however, different line representation can be used for each separate curve. The line weight of curves should be heavier than the grid ruling.

A key, or legend, should be placed in an available portion of the grid, preferably enclosed in a border, to define point symbols or line types that are used for curves. Remember that the lines representing each curve should never be drawn through the symbols, so that the precise point is always identifiable. Figure 2.21 demonstrates the use of a key and the practice of breaking the line at each symbol.

2.3.7
Titles

Each graph must be identified with a complete title. The title should include a clear, concise statement of the data being represented, along with items such as the name of the author, the data of the experiment, and any and all information concerning

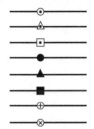

Figure 2.19
Symbols.

Figure 2.20
Line representation.

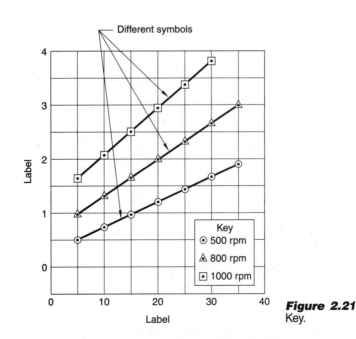

Figure 2.21
Key.

the plot, including the name of the institution or company. Titles are normally enclosed in a border.

All lettering, the axes, and the curves should be sufficiently bold to stand out on the graph paper. Letters should be neat and of standard size. Figure 2.22 is an illustration of plotted experimental data incorporating many of the items discussed in the chapter.

2.3.8
Computer-Assisted Plotting

A number of commercial software packages are available to produce graphs. The quality and accuracy of these computer-generated graphs vary, depending on the sophistication of the software as well as on the plotter or printer employed. Typically, the software will produce an axis scale graduated and calibrated to accommodate the range of data values that will fit the paper. This may or may not produce a readable or interpretable scale. Therefore, it is necessary to apply considerable judgment depending on the results needed. For example, if the default plot does not meet needed scale readability, it may be necessary to specify the scale range to achieve an appropriate scale graduation, since this option allows greater control of the scale drawn.

Figure 2.22
Sample plot.

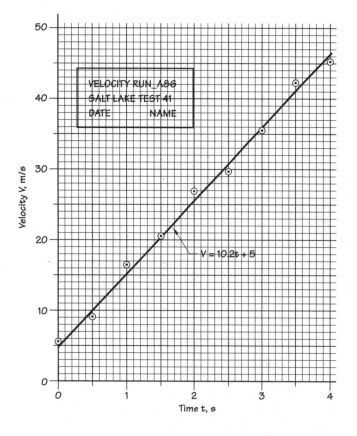

Necessary steps to follow when plotting a graph.

1. Select the type of graph paper (rectilinear, semilog, log-log, etc.) and grid spacing for best representation of the given data.

2. Choose the proper location of the horizontal and vertical axes.

3. Determine the scale units (range) for each axis to display the data appropriately.

4. Graduate and calibrate the axes (1, 2, 5 rule).

5. Identify each axis completely.

6. Plot points and use permissible symbols.

7. Draw the curve or curves.

8. Identify each curve and add title and necessary notes.

9. Darken lines for good reproduction.

Computer-produced graphics with uniform scales may not follow the 1, 2, 5 rule, particularly since the software plots the independent variable based on the data collected. If the software has the option of separately specifying the range, that is, plotting the data as an *X-Y* scatter plot, you will be able to achieve scale graduations and calibrations that do follow the 1, 2, 5 rule, making it easier to read values from the graph. The hand-plotted graph that was illustrated in Fig. 2.22 is plotted using EXCEL. The results are shown with the default plot in Fig. 2.23a and using the *X-Y* scatter plot with a linear curve-fit in Fig. 2.23b.

Process

1. Record via keyboard or import data into spreadsheet.
2. Select independent (*x*-axis) and dependent variable(s).
3. Select appropriate graph (style or type) from menu.
4. Produce trial plot with default parameters.
5. Examine (modify as necessary) origin, range, graduation, and calibrations. Note: use the 1, 2, 5 rule.
6. Label each axis completely.
7. Select appropriate plotting-point symbols and legend.
8. Create complete title.
9. Examine plot and store the data.
10. Plot or print the data.

2.4

Empirical Functions

Empirical functions are generally described as those based on values obtained by experimentation. Since they are arrived at experimentally, equations normally available from theoretical derivations are not possible. However, mathematical expressions can be modeled to fit experimental functions, and it is possible to classify most empirical results into one of four general categories: (1) linear, (2) exponential, (3) power, or (4) periodic.

A linear function, as the name suggests, will plot as a straight line on uniform rectangular coordinate paper. Likewise, when a curve representing experimental data is a straight line or a close approximation to a straight line, the relationship of the variables can be expressed by a linear equation.

Correspondingly, exponential equations, when plotted on semilog paper, will be linear. The basic form of the equation is $y = be^{mx}$. This can be written in log form and becomes $\log y = mx \log e + \log b$. Alternatively, using natural logarithms, the equation becomes $\ln y = mx + \ln b$ because $\ln e = 1$. The independent variable x is plotted against $\ln y$.

The power equation has the form of $y = bx^m$. Written in log form it becomes $\log y = m \log x + \log b$. This data will plot straight on log-log paper, since the log of the independent variable x is plotted against the log of y.

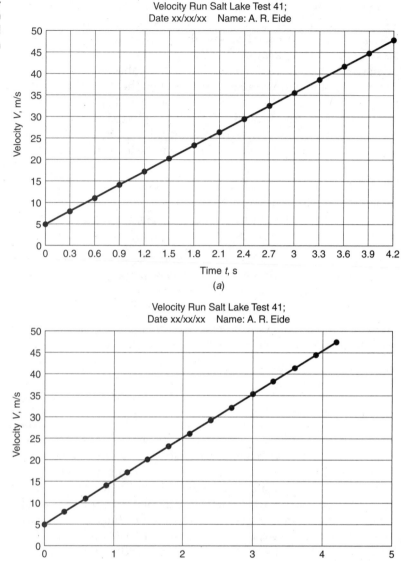

Figure 2.23

When the data represent experimental results and a series of points are plotted to represent the relationship between the variables, it is improbable that a straight line can be constructed through every point, since some error is inevitable. If all points do not lie on the same line, an approximation scheme or averaging method may be used to arrive at the best possible fit.

2.5

Curve Fitting

Different methods or techniques are available to arrive at the best "straight-line" fit. Three methods commonly employed for finding the best fit are as follows:

1. Method of selected points
2. Method of averages
3. Method least squares

Each of these techniques is progressively more accurate. The first method will be briefly described in Sec. 2.6. Several examples will be presented in this chapter to demonstrate correct methods for the representation of technical data.

Method 2, the method of averages, is based on the idea that the line location is positioned to make the algebraic sum of the absolute values of the differences between observed and calculated values of the ordinate equal to 0.

In both methods 2 and 3, the procedure involves minimizing what are called *residuals,* or the difference between an observed ordinate and the corresponding computed ordinate. The method of averages will not be applied in this book, but there are any number of reference texts available that adequately cover the concept.

2.6

Method of Selected Points

The method of selected points is a valid method of determining the equation that best fits data that exhibit a linear relationship. Once the data have been plotted and determined to be linear, a line is selected that appears to fit the data best. This is most often accomplished by visually selecting a line that goes through as many data points as possible and has approximately the same number of data points on either side of the line.

Once the line has been constructed, two points, such as A and B, are selected *on the line* and at a reasonable distance apart. The coordinates of both points $A(X_1,Y_1)$ and $B(X_2,Y_2)$ must satisfy the equation of the line, since both are points on the line.

2.7

Empirical Equations—Linear

When experimental data plot as a straight line on rectangular grid paper, the equation of the line belongs to a family of curves whose basic equation is given by

$$y = mx + b \tag{2.1}$$

where m is the slope of the line, a constant, and b is a constant referred to as the *y intercept* (the value of y when $x = 0$).

To demonstrate how the method of selected points works, consider the following example.

Example problem 2.1   The velocity V of an automobile is measured at specified time t intervals. Determine the equation of a straight line constructed through the points recorded in Tab. 2.3. Once an analytic equation has been determined, velocities at intermediate values can be computed.

Table 2.3

Time t, s	0	5	10	15	20	25	30	35	40
Velocity V, m/s	24	33	62	77	105	123	151	170	188

Procedure

1. Plot the data on rectangular paper or do a computer scatter-plot. If the results form a straight line (see Fig. 2.24), the function is linear and the general equation is of the form

$$V = mt + b$$

where m and b are constants.

2. Select two points on the line, $A(t_1, V_1)$ and $B(t_2, V_2)$, and record the value of these points. Points A and B should be widely separated to reduce the effect on m and b of errors in reading values from the graph. Points A and B are identified on Fig. 2.24 for instructional reasons. They should not be shown on a completed graph that is to be displayed.

$A(10, 60)$
$B(35, 165)$

Figure 2.24
Data plot.

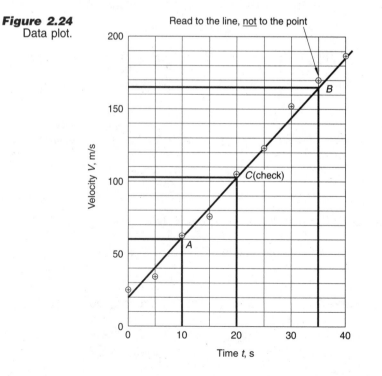

3. Substitute the points *A* and *B* into $V = mt + b$.

$60 = m(10) + b$

$165 = m(35) + b$

4. The equations are solved simultaneously for the two unknowns.

$m = 4.2$

$b = 18$

5. The general equation of the line can be written as

$V = 4.2t + 18$

6. Using another point $C(t_3, V_3)$, check for verification:

$C(20, 102)$

$102 = 4.2(20) + 18$

$102 = 84 + 18 = 102$

It is also possible by discreet selection of points to simplify the solution. For example, if point *A* is selected at (0, 18) and this coordinate is substituted into the general equation

$18 = m(0) + b$ Note: Approximately 20 on graph

the constant *b* is immediately known.

Empirical Equations—Power Curves

When experimentally collected data are plotted on rectangular coordinate graph paper and the points do not form a straight line, you must next determine which family of curves the line most closely approximates. Consider the following familiar example.

Example problem 2.2   Suppose that a solid object is dropped from a tall building, and the values are as recorded in Tab. 2.4.

Solution   To anyone who has studied fundamental physics, it is apparent that these values should correspond to the general equation for a free-falling body (neglecting air friction):

$s = \frac{1}{2}gt^2$

But assume for a moment that all we have is the table of values.

First it is helpful to make a freehand plot to observe the data visually (see Fig. 2.25). From this quick plot, the data points are more easily recognized as belonging to a family of curves whose general equation can be written

$y = bx^m$ (2.2)

Table 2.4

Time *t*, s	Distance *s*, m/s
0	0
1	4.9
2	19.6
3	44.1
4	78.4
5	122.5
6	176.4

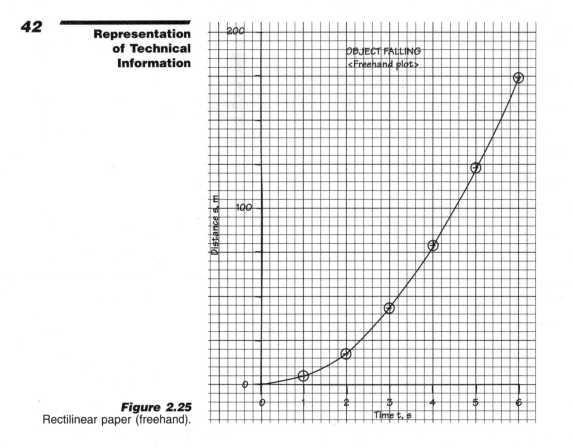

Figure 2.25
Rectilinear paper (freehand).

Remember that before the method of selected points can be applied to determine the equation of the line in this example problem, the plotted line must be straight, because two points on a curved line do not uniquely identify the line. Mathematically, this general equation can be modified by taking the logarithm of both sides,

$\log y = m \log x + \log b$, or

$\ln y = m \ln x + \ln b$

This equation suggests that if the logs of both y and x were determined and the results plotted on rectangular paper, the line would likely be straight.

Realizing that the log of 0 is undefined and plotting the remaining points that are recorded in Tab. 2.5 for Log s versus Log t, the results are shown in Fig. 2.26.

Since the graph of log s versus log t does plot as a straight line, it is now possible to use the general form of the equation

$\log y = m \log x + \log b$

and apply the method of selected points.

Table 2.5

Time t, s	Distance s, m	Log t	Log s
0	0		
1	4.9	0.000 0	0.690 2
2	19.6	0.301 0	1.292 3
3	44.1	0.477 1	1.644 4
4	78.4	0.602 1	1.894 3
5	122.5	0.699 0	2.088 1
6	176.4	0.778 2	2.246 5

When reading values for points A and B from the graph, we must remember that the logarithm of each variable has already been determined and the values plotted.

$A(0.2, 1.09)$

$B(0.6, 1.89)$

Points A and B can now be substituted into the general equation $\log s = m \log t + \log b$ and solved simultaneously.

$1.89 = m(0.6) + \log b$

$1.09 = m(0.2) + \log b$

$m = 2.0$

$\log b = 0.69$

$b = 4.9$

As examination of Fig. 2.26 shows that the value of $\log b$ (0.69) can be read from the graph where $\log t = 0$. This, of course, is where $t = 1$ and is the y intercept for log-log plots.

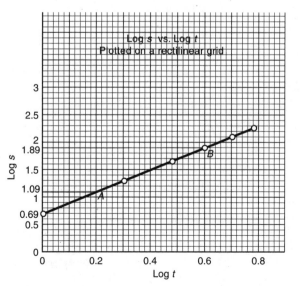

Figure 2.26
Log-log on rectilinear grid paper.

The general equation can then be written as

$$s = 4.9t^{2.0}$$

Or,

$$s = \tfrac{1}{2}gt^2,$$

where $g = 9.8 \ m/s^2$.

One obvious inconvenience is the necessity of finding logarithms of each variable and then plotting the logs of these variables.

This step is not necessary since functional paper is commercially available with $\log x$ and $\log y$ scales already constructed. Log-log paper allows the variables themselves to be plotted directly without the need of computing the log of each value.

In the preceding example, once the general form of the equation is determined (Eq. [2.2]), the data can be plotted directly on log-log paper. Since the resulting curve is a straight line, the method of selected points can be used directly (see Fig. 2.27).

The log form of the equation is again used:

$$\log s = m \log t + \log b$$

Select points A and B on the line:

$A(1.5, 11)$

$B(6, 175)$

Figure 2.27
Log-log paper.

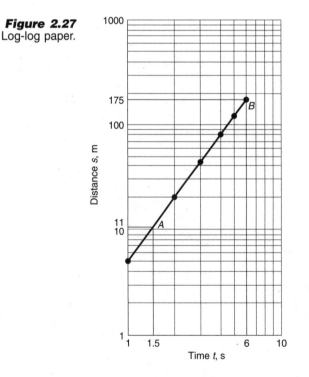

Substitute the values into the general equation $\log s = m \log t + \log b$, taking careful note that the numbers are the variables and *not* the logs of the variables.

$\log 175 = m \log 6 + \log b$

$\log 11 = m \log 1.5 + \log b$

Again, solving these two equations simultaneously results in the following approximate values for the constants b and m:

$b = 4.897\ 8 = 4.9$

$m = 1.995\ 7 = 2.0$

Identical conclusions can be reached:

$s = \frac{1}{2}gt^2$

This time, however, one can use functional scales rather than calculate the log of each number.

Suppose your data do not plot as a straight line (or nearly straight) on rectangular coordinate paper nor is the line approximately straight on log-log paper. Without experience in analyzing experimental data, you may feel lost as to how to proceed. Normally, when experiments are conducted you have an idea as to how the parameters are related and you are merely trying to quantify that relationship. If you plot your data on semilog graph paper and it produces a reasonably straight line, then it has the general form $y = be^{mx}$.

Empirical Equations— Exponential Curves

Example problem 2.3   Assume that an experiment produces the data shown in Tab. 2.6.

Solution   The data (Tab. 2.6) when plotted produce the graph shown as Fig. 2.28. To determine the constants in the equation $y = be^{mx}$, write it in linear form, either as

$\log y = mx \log e + \log b$

or

$\ln y = mx + \ln b$

The method of selected points can now be employed for $\ln FC = mV + \ln b$ (choosing the natural log form). Points $A(15, 33)$ and $B(65, 470)$ are carefully selected on the line, so they must satisfy the equation. Substituting the values of V and FC at points A and B, we get

$\ln 470 = 65m + \ln b$

and

$\ln 33 = 15m + \ln b$

Table 2.6

Velocity V, m/s	Fuel Consumption FC, mm^3/s
10	25.2
20	44.6
30	71.7
40	115
50	202
60	367
70	608

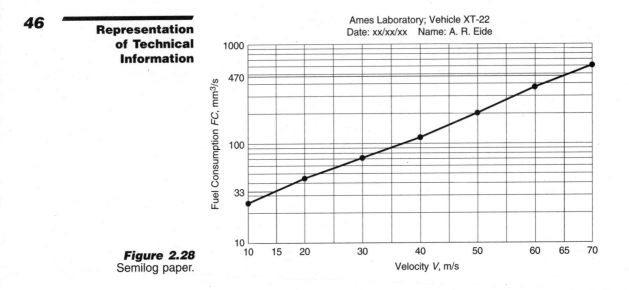

Figure 2.28
Semilog paper.

Solving simultaneously for *m* and *b*, we have

$m = 0.053\,1$

and

$b = 14.9$

The desired equation then is determined to be $FC = 15e^{(0.05V)}$. This determination can be checked by choosing a third point, substituting the value for *V*, and solving for *FC*.

The following terms are basic to the material in Chapter 2. You should be able to define these terms and be able to interpret them into various applications.

2.10

Key Terms and Concepts

Scale graduations
Scale calibrations
Uniform scales
Nonuniform scales
1, 2, 5 rule
Plotting procedures
Graph symbols
Axis identification

Line representation
Titles
Empirical functions
Curve fitting
Method of selected points
Linear curves
Power curves
Exponential curves

2.1 Table 2.7 is data from trial run on the Utah salt flats made by an experimental turbine-powered vehicle.

(a) Plot the data on rectilinear paper using time as the independent variable.

(b) Determine the equation of the line, using the method of selected points.

(c) Interpret the slope of the line.

2.2 Table 2.8 lists the values of velocity recorded on a ski jump in Colorado this past winter.

(a) Plot the data on rectilinear paper using time as the independent variable.

(b) Using the method of selected points, determine the equation of the line.

(c) Give the average acceleration.

2.3 Table 2.9 is a collection of data for iron-constantan thermocouple. Temperature is in degrees Celsius and the electromotive force (emf) is in millivolts.

(a) Plot a graph, using rectilinear paper, showing the relation of temperature to voltage, with voltage as the independent variable.

(b) Using the method of selected points, find the equation of the line.

Table 2.7

Time t, s	Velocity V, m/s
10.0	15.1
20.0	32.2
30.0	63.4
40.0	84.5
50.0	118
60.0	139

Table 2.8

Time t, s	Velocity V, m/s
1.0	5.3
4.0	18.1
7.0	26.9
10.0	37.0
14.0	55.2

Table 2.9

Voltage (emf), mV	Temperature t, °C
2.6	50.0
6.7	100.0
8.8	150.0
11.2	200.0
17.0	300.0
22.5	400.0
26.0	500.0
32.5	600.0
37.7	700.0
41.0	800.0
48.0	900.0
55.2	1 000.0

2.4 A Sessions pump was tested to determine the power required to produce a range of discharges. The test was performed in Oak Park on July 1, 1984. The results of the test are shown in Tab. 2.10.

(a) Plot a graph showing the relation of power required to discharge.

(b) Determine the equation of the relationship.

(c) Predict the power required to produce a discharge of 37 L/s.

Table 2.10

Discharge Q, L/s	Power P, kW
3.00	28.5
7.00	33.8
10.00	39.1
13.50	43.2
17.00	48.0
20.00	51.8
25.00	60.0

Table 2.11

Deflection D, mm	Load L, N
2.25	35.0
12.0	80.0
20.0	120.0
28.0	160.0
35.0	200.0
45.0	250.0
55.0	300.0

2.5 A spring was tested at AMAC Manufacturing. The test of spring ZX-15 produced the data shown in Tab. 2.11.

(a) Plot the data on rectangular graph paper and determine the equation that expresses the deflection to be expected under a given load.

(b) Predict the load required to produce a deflection of 30 mm.

(c) What deflection would be expected to produce a load of 50 N?

2.6 A Johnson furnace was tested in northern Minnesota to determine the heat generated, expressed in thousands of British thermal units per cubic foot of furnace volume, at varying temperatures. The results are shown in Tab. 2.12.

(a) Plot data on rectilinear paper, with temperature as the independent variable.

(b) Plot the data on log-log graph paper, with temperature as the independent variable.

(c) Using the method of selected points, determine the equation that best fits the data.

(d) What would be the heat released for a temperature of 1000°F?

Table 2.12

Heat released H, 10^3 Btu/ft^3	Temperature T, °F
0.200	172
0.600	241
2.00	392
4.00	483
8.00	608
20.00	812
40.00	959
80.00	1 305

2.7 The capacity of a 20-cm screw conveyor that is moving dry ground corn is expressed in liters per second and the conveyor speed in revolutions per minute. A test was conducted in Cleveland on conveyor model JD172 last week. The results of the test are given in Tab. 2.13.

(a) Plot the data on rectilinear graph paper.

(b) Plot the data on semilog paper.

(c) Plot the data on log-log paper.

(d) Determine the equation that expresses velocity in terms of capacity utilizing the proper plot.

Table 2.13

Capacity C, L/s	Angular velocity V, r/min
3.01	10.0
6.07	21.0
15.0	58.2
30.0	140.6
50.0	245
80.0	410
110.0	521

2.8 The resistance of a class and shape of electrical conductor was tested over a wide range of sizes at constant temperature. The test was performed in Madison, Wisconsin, on April 4, 1984, at the Acme Electrical Labs. The test results are shown in Tab. 2.14. The resistance is expressed in milliohms per meter of conductor length.

- (a) Plot the data on log-log paper.
- (b) Using the method of selected points, find the equation that expresses resistance as a function of the area of the conductors.

2.9 The area of a circle can be expressed by the formula $A = \pi R^2$. If the radius varies from 0.5 to 5 cm, perform the following:

- (a) Construct a table of radius versus area mathematically. Use radius increments of 0.5 cm.
- (b) Construct a second table of log R versus log A.
- (c) Plot the values from (a) on log-log paper and determine the equation of the line.
- (d) Plot the values from (b) on rectilinear paper and determine the equation of the line.

2.10 The volume of a sphere is $V = \frac{4}{3}\pi R^3$.

- (a) Prepare a table of volume versus radius, allowing the independent variable radius to vary from 2.0 to 10.0 m in 1-m increments.
- (b) Plot a graph on log-log paper showing the relation of volume to radius using the values from the table in (a).
- (c) Verify the equation, $V = \frac{4}{3}\pi R^3$, by the method of selected points.

2.11 A 90° triangular weir is commonly used to measure flow rate in a stream. Data on the discharge through the weir were collected and recorded as shown in Tab. 2.15.

- (a) Plot the data on log-log paper, with height as the independent variable.
- (b) Determine the equation of the line using the method of selected points.

Table 2.14

Area A, mm²	Resistance R, mΩ/m
0.021	505
0.062	182
0.202	55.3
0.523	22.2
1.008	11.3
3.32	4.17
7.29	1.75

Table 2.15

Discharge, Q, m³/s	1.5	8	22	45	78	124	182	254
Height, h, m	1	2	3	4	5	6	7	8

2.12 A pitot tube is a device for measuring the velocity of flow of a fluid (see Fig. 2.29). A stagnation point occurs at point 2; by recording the height differential h, the velocity at point 1 can be calculated. Assume for this problem that the velocity at point 1 is known corresponding to the height differential h. Table 2.16 records these values.

- (a) Plot the data on log-log paper using height as the independent variable.
- (b) Determine the equation of the line using the method of selected points.

Table 2.16

Velocity V, m/s	1.4	2.0	2.8	3.4	4.0	4.4
Height h, m	0.1	0.2	0.4	0.6	0.8	1.0

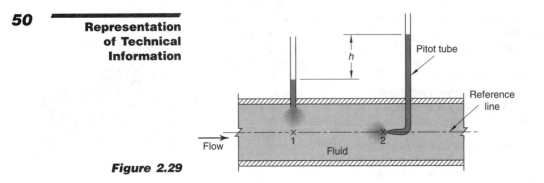

Figure 2.29

2.13 A new production facility manufactured 29 parts the first month, but then increased production, as shown in Tab. 2.17.
    (*a*) Plot the data on semilog paper.
    (*b*) Using the time variable that defines January to be 1, February to be 2, and so on, determine the equation.

Table 2.17

Month	Jan	Feb	Mar	Apr	May	Jun	Jul	Aug
Number	29	40	48	58	85	115	124	180

Table 2.18

Time t, s	Voltage V, V
6	98
10	62
17	23
25	9.5
32	3.5
38	1.9
42	1.33

2.14 The voltage across a capacitor during discharge was recorded as a function of time (see Tab. 2.18).
    (*a*) Plot the data on semilog paper, with time as the independent variable.
    (*b*) Determine the equation of the line using the method of selected points.

2.15 When a capacitor is to be discharged, the current flows until the voltage across the capacitor is 0. This current flow when measured as a function of time resulted in the data given in Tab. 2.19.
    (*a*) Plot the data points on semilog paper, with time as the independent variable.
    (*b*) Determine the equation of the line using the method of selected points.

Table 2.19

Current I, A	1.81	1.64	1.48	1.34	1.21	0.73
Time t, s	0.1	0.2	0.3	0.4	0.5	1.0

2.16 When fluid is flowing in the line, it is relatively easy to begin closing a valve that is wide open. But as the valve approaches a more nearly closed position, it becomes considerably more difficult to force movement. Visualize a circular pipe with a simple flap hinged at one edge being closed over the end of the pipe. The fully open position is $\theta = 0$, and the fully closed condition is $\theta = 90°$ (see Fig. 2.30).

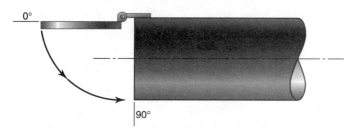

Figure 2.30

A test was conducted on such a valve by applying a constant torque at the hinge position and measuring the angular movement of the valve. The test data are shown in Tab. 2.20.

(a) Plot the data, with angle as the independent variable.

(b) By the method of selected points, find the equation relating torque to angular movement.

2.17 All materials are elastic to some extent. It is desirable that certain parts of some designs compress when a load is applied to assist in making the part airtight or watertight (such as a jar lid). The test results shown in Tab. 2.21 resulted from a test made at the Herndon Test Labs in Houston on a material known as SILON Q-177.

(a) Plot the data on semilog graph paper.

(b) Using the method of selected points, find the equation of the relationship.

(c) What pressure would cause a 10 percent compression?

2.18 The rate of absorption of radiation by metal plates varies with the plate thickness and the nature of the source of radiation. A test was made at Ames Labs on October 11, 1982, using a Geiger counter and a constant source of radiation; the results are shown in Tab. 2.22.

(a) Plot the data on semilog graph paper.

(b) Find the equation of the relationship between the parameters.

(c) What level of radiation would you estimate to pass a 2-in-thick plate of the metal used in the test described above?

Table 2.20

Torque T, N·m	Movement θ, degrees
3.0	5.2
6.0	29.3
10.0	40.9
20.0	56.3
35.0	71.0
50.0	84.8

Table 2.21

Pressure P, Mpa	Relative compression R, %
1.12	27.3
3.08	37.6
5.25	46.0
8.75	50.6
12.3	56.1
16.1	59.2
30.2	65.0

Table 2.22

Plate thickness W, mm	Geiger counter C, counts per second
0.20	5 500
5.00	3 720
10.00	2 550
20.00	1 320
27.5	720
32.5	480

Engineering Estimations and Approximations

3

Within each engineering discipline there are specialty areas which give the appearance that engineering is a diverse profession with little commonality in the tasks performed. However, engineers are problem solvers, creating new designs which satisfy a need and improve the living standard. During the design process, engineers of all specialists will need to acquire physical measurements pertaining to the product or system being designed, the environment in which the design will operate, or both.

The nineteenth century physicist Lord Kelvin stated that man's knowledge and understanding are not of high quality unless the information can be expressed in numbers. We all have made or heard statements such as "the water is too hot." This statement may or may not give us an indication of the temperature of the water. At a given temperature water may be too hot for taking a bath but not hot enough for making instant coffee or tea.

The truth is that pronouncements such as "hot," "too hot," "not very hot," and so on, are relative to a standard selected by the speaker and have meaning only to those who know what that standard is.

Engineers make measurements of a vast array of physical quantities that control the design solution. Skill in making and interpreting measurements is an essential element in our practice of engineering.

Any physical measurement cannot be assumed to be exact. Errors are likely to be present regardless of the precautions used when making the measurement. Quantities determined by analytical means are not always exact either. Often assumptions are made to arrive at an analytical expression which is then used to calculate a numerical value.

53

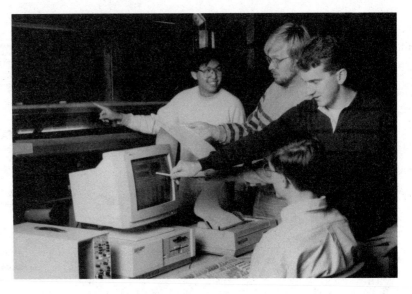

Figure 3.1
Measurements from a wind tunnel experiment are acquired electronically and immediately analyzed with software. The test can be quickly modified as needed. *(Iowa State University.)*

Table 3.1

Quantity	Number of significant figures
4 784	4
36	2
60	1 or 2
600	1, 2, or 3
6.00×10^2	3
31.72	4
30.02	4
46.0	3
0.02	1
0.020	2
600.00	5

It is clear that a method of expressing results and measurements is needed that will convey how "good" these numbers are. The use of significant digits gives us this capability without resorting to the more rigorous approach of computing an estimated percentage error to be specified with each numerical result or measurement.

A *significant digit,* or *figure,* is defined as any digit used in writing a number, *except* those zeros that are used only for location of the decimal point or those zeros that do not have any nonzero digit on their left. When you read the number 0.001 5, only the digits 1 and 5 are significant, since the three zeros have no nonzero digit to their left. We would say then that this number has two significant figures. If the number is written 0.001 50, it contains three significant figures; the rightmost zero is significant.

Numbers 10 or larger that are not written in scientific notation and that are not counts (exact values) can cause difficulties in interpretation when zeros are present. For example, 2 000 could contain one, two, three, or four significant digits; it is not clear which. If you write the number in scientific notation as 2.000×10^3, then clearly four significant digits are intended. If you want to show only two significant digits, you would write 2.0×10^3. It is our recommendation that, if uncertainty results from using standard decimal notation, you switch to scientific notation so your reader can clearly understand your intent. Table 3.1 shows the number of significant figures for several quantities.

You may find yourself as the user of values where the writer was not careful to properly show significant figures. What then? Assuming that the number is not a count or a known exact value, about all you can do is establish a reasonable number of significant figures based on the context of the value and on your experience. Once you have decided on a reasonable number of significant digits, you can then use the number in any calculations that are required.

When reading instruments, such as an engineer's scale, thermometer, or fuel gauge, the last digit will normally be an estimate. That is, the instrument is read by estimating between the smallest graduations on the scale to get the final digit. In Fig. 3.2a, you may estimate the reading from the engineer's scale to be 1.27, with the 7 being a doubtful digit of the three significant figures. It is standard practice to count the doubtful digit as significant, thus the 1.27 reading has three significant figures. Similarly, the thermometer in Fig. 3.2b may be read as 52.8° with the 8 being doubtful.

In Fig. 3.2c, the graduations create a more difficult task for reading a fuel level. Each graduation is one-sixth of a full tank. The reading appears to be about three-fourths of the distance between one-sixth and two-sixths making the reading seven twenty-fourths of a tank or 0.292. How many significant figures are there? In this case one significant figure is all that can be obtained, so the answer should be rounded to 0.3. The difficulty in this example is not the significant figures, but the scale of the fuel gauge. It is meant to convey a general impression of the fuel level and not a numerically significant value. Furthermore, the automobile manufacturer did not deem that the cost of a more accurate and precise fuel-measurement system was justified. Therefore, the selection of the instrument is an important factor in physical measurements.

Calculators and computers commonly work with numbers having as few as 7 digits or as many as 16 or 17 digits. This is true no matter how many significant digits an input value or calculated value should have. Therefore, you will need to exercise care in reporting values from a calculator display or from a computer output. Most high-level computer languages allow you to control the number of digits that are to be displayed or printed. If a computer output is to be a part of your final solution presentation, you will need to carefully control the output form. If the output is only an intermediate step, you can round the results to a reasonable number of significant figures in your presentation.

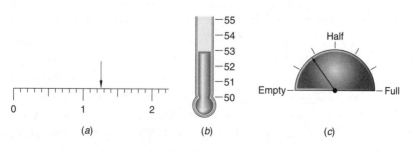

Figure 3.2
Reading graduations on instruments will include a doubtful, or estimated, value.

As you perform arithmetic operations, it is important that you not lose the significance of your measurements or, conversely, imply precision that does not exist. Rules for determining the number of significant figures that should be reported following computations have been developed by engineering associations. The following rules customarily apply.

1. Multiplication and Division

The product or quotient should contain the same number of significant digits as are contained in the number with the fewest significant digits.

Examples

a. (2.43)(17.675) = 42.950 25

If each number in the product is exact, the answer should be reported as 42.950 25. If the numbers are not exact, as is normally the case, 2.43 has three significant figures and 17.675 has five. Applying the rule, the answer should contain three significant figures and be reported as 43.0 or 4.30×10^1.

b. (2.479 h)(60 min/h) = 148.74 min

In this case, the conversion factor is exact (a definition) and could be thought of as having an infinite number of significant figures. Thus, 2.479, which has four significant figures, controls the precision, and the answer is 148.7 min, or 1.487×10^2 min.

c. $(4.00 \times 10^2$ kg)(2.204 6 lbm/kg) = 881.84 lbm

Here, the conversion factor is not exact, but you should not let the conversion factor dictate the precision of the answer if it can be avoided. You should attempt to maintain the precision of the value being converted; you cannot improve its precision. Therefore, you should use a conversion factor that has one or two more significant figures than will be reported in the answer. In this situation, three significant figures should be reported, yielding 882 lbm.

d. 589.62/1.246 = 473.210 27

The answer, to four significant figures, is 473.2.

2. Addition and Subtraction

The answer should show significant digits only as far to the right as is seen in the least precise number in the calculation.

Examples

a. 1 725.463
   189.2
    16.73

1 931.393

The least precise number in this group is 189.2 so, according to the rule, the answer should be reported as 1 931.4. Using alternative reasoning, suppose these numbers are instrument readings, which means the last reported number in each is a doubtful digit. A column addition that contains a doubtful digit will result in a doubtful digit in the sum. So all three digits to the right of the decimal in the answer are doubtful. Normally we report only one; thus the answer is 1 931.4 after rounding.

b. 897.0
  $- 0.092\ 2$

  896.907 8

Application of the rule results in an answer of 896.9.

3. Combined Operations

If products or quotients are to be added or subtracted, perform the multiplication or division first, establish the correct number of significant figures in the subanswer, perform the addition or subtraction, and round to proper significant figures. Note, however, that in calculator or computer applications it is not practical to perform intermediate rounding. It is normal practice to perform the entire calculation and then report a reasonable number of significant figures.

If results from additions or subtractions are to be multiplied or divided, an intermediate determination of significant figures can be made when the calculations are performed manually. Use the suggestion already mentioned for calculator or computer answers.

Subtractions that occur in the denominator of a quotient can be a particular problem when the numbers to be subtracted are very nearly the same. For example, $39.7/(772.3 - 772.26)$ gives 992.5 if intermediate roundoff is not done. If, however, the subtraction in the denominator is reported with one digit to the right of the decimal, the denominator becomes zero and the result becomes undefined. Commonsense application of the rules is necessary to avoid problems.

4. Rounding

In rounding a value to the proper number of significant figures, *increase the last digit retained by 1 if the first figure dropped is 5 or greater.* This is the rule normally built into a calculator display control or a control language.

Examples

a. 827.48 rounds to 827.5 or 827 for four and three significant digits, respectively.
b. 23.650 rounds to 23.7 for three significant figures.
c. 0.014 3 rounds to 0.014 for two significant figures.

In measurements, accuracy and precision have different meanings and cannot be used interchangeably. *Accuracy* is a measure of the nearness of a value to the correct or true value. *Precision* refers to the repeatability of a measurement, that is, how close successive measurements are to each other. Figure 3.3 illustrates accuracy and precision of the results of four dart throwers. Thrower a is both inaccurate and imprecise because the results are away from the bullseye (accuracy) and widely scattered (precision). Thrower b is accurate because the throws are evenly distributed about the desired result but imprecise because of the wide scatter. Thrower c is precise with the tight cluster of throws but inaccurate because the results are away from the desired bullseye. Finally, thrower d demonstrates accuracy and precision with tight cluster of throws around the center of the target. Throwers a, b, and c can improve their performance by analyzing the causes for the errors. Body position, arm motion, and release point could cause deviation from the desired result.

Engineers making physical measurements encounter two types of errors, systematic and random. These will be discussed in the next section.

Measurements can be reported as a value plus or minus ($\pm$) a number, for example, 32.3 ± 0.2. This indicates a range of values which are equally representative of the indicated value (32.3). Thus 32.3, 32.1, and 32.5 are among the "acceptable" values for this measurement. A range of permissible error can also be specified as a percentage of the indicated value. For example, a ther-

Figure 3.3
Illustration of the difference between accuracy and precision in physical measurements.

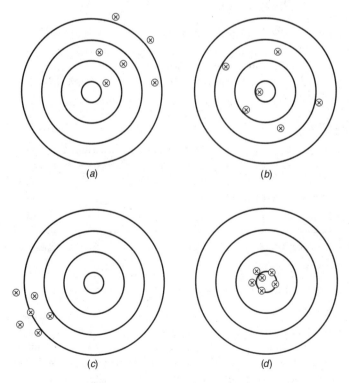

mometer's accuracy may be specified as ± 1.0 percent of full scale reading. Thus if the full scale reading is 220°F, readings should be within ± 2.2 of the true value ($220 \times 0.01 = 2.2$).

Errors

To measure is to err! Any time a measurement is taken, the result is being compared to a true value, which itself may not be known exactly. If we measure the dimensions of a room, why doesn't a repeat of the measurements yield the same results? Did the same person make all measurements? Was the same measuring instrument used? Were the readings all made from exactly the same eye position? Was the measuring instrument correctly graduated? It is obvious that errors will occur in each measurement. We must try to identify the errors if we can and correct them in our results. If we can't identify the error, we must provide some conclusions as to the resulting accuracy and precision of our measurements.

Identifiable and correctable errors are classified as systematic; accidental or other nonidentifiable errors are classified as random.

3.4.1
Systematic Errors

Our task is to measure the distance between two fixed points. Assume that the distance is in the range of 1 200 m and that we are experienced and competent and have equipment of high quality to do the measurement. Some of the errors that occur will always have the same sign ($+$ or $-$) and are said to be systematic. Assume that a 25 m steel tape is to be used, one that has been compared with the standard at the U.S. Bureau of Standards in Washington, D.C. If the tape is not exactly 25.000 m long, then there will be a systematic error each of the 48 times that we use the tape to measure out the 1 200 m.

However, the error can be removed by applying a correction. A second source of error can stem from a difference between the temperature at the time of use and at the time when the tape was compared with the standard. Such an error can be removed if we measure the temperature of the tape and apply a mathematical correction. The coefficient of thermal expansion for steel is 11.7×1.0^{-6} per kelvin). The accuracy of such a correction depends on the accuracy of the thermometer and on our ability to measure the temperature of the tape instead of the temperature of the surrounding air. Another source of systematic error can be found in the difference in the tension applied to the tape while in use and the tension employed during standardization. Again, scales can be used but, as before, their accuracy will be suspect. In all probability, the tape was standardized by laying it on a smooth surface and supporting it throughout. But such surfaces are seldom available in the field. The tape is suspended at times, at least partially. But, knowing the weight of the tape, the tension that is applied, and the length of the suspended tape, we can calculate a correction and apply it.

The sources of systematic error just discussed are not all the possible sources, but they illustrate an important problem even encountered in taking comparatively simple measurements. Similar problems occur in all types of measurements: mechanical quantities, electrical quantities, mass, sound, odors, and so forth. We must be aware of the presence of systematic errors, eliminate those that we can, and quantify and correct for those remaining.

3.4.2
Random Errors

In reading Sec. 3.4.1 you may have realized that even if it had been possible to eliminate all the systematic errors, the measurement is still not exact. To elaborate on this point, we will continue with the example of the task of measuring the 1 200 m distance. Several random errors can creep in, as follows. When reading the thermometer, we must estimate the reading when the indicator falls between graduations. Moreover, it may appear that the reading is exactly on a graduation when it is actually slightly above or below the graduation. Furthermore, the thermometer may not be accurately measuring the tape temperature but may be influenced instead by the temperature of the ambient air. These errors can thus produce measurements that are either too large or too small. Regarding sign and magnitude, the error is therefore random.

Errors can also result from our correcting for the sag in a suspended tape. In such a correction, it is necessary to determine the weight of the tape, its cross-sectional area, its modulus of elasticity, and the applied tension. In all such cases, the construction of the instruments used for acquiring these quantities can be a source of both systematic and random errors.

The major difficulty we encounter with respect to random errors is that, although their presence is obvious by the scatter in the data, it is impossible to predict the magnitude and sign of the accidental error that is present in any one measurement. Repeating measurements and averaging the results will reduce the random error in the average. However, repeating measurements will not reduce the systematic error in the average result.

Refinement of the apparatus and care in its use can reduce the magnitude of the error; indeed, many engineers have devoted their careers to this task.

Likewise, awareness of the problem, knowledge about the degree of precision of the equipment, skill with measurement procedures, and proficiency in the use of statistics allow us to determine the approximate magnitude of the error remaining in

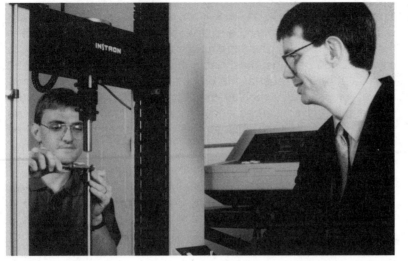

measurements. This knowledge, in turn, allows us to accept the error or develop different apparatus and/or methods in our work. It is beyond the scope of this text to discuss quantifying accidental errors.

3.5

Approximations

Engineers strive for a high level of precision in their work. However, it is also important to be aware of the expected precision and the time and cost of attaining it. There are many instances where an engineer is expected to make an approximation to an answer, that is estimate the result with reasonable accuracy but under tight time and cost constraints. To do this engineers rely on their basic understanding of the problem under discussion coupled with their previous experience. This knowledge and experience is what distinguishes an "approximation" from a "guess." If greater accuracy is needed, the initial approximation can be refined when time and funds are available and the necessary data for refining the result are available.

In the area of our highest competency, we are expected to be able to make rough estimates to provide figures that can be used for tentative decisions. These estimates may be in error by perhaps 10 to 20 percent or even more. The accuracy of these estimates depends strongly on what reference materials we have available, how much time is allotted for the estimate, and, of course, how experienced we are with similar problems. The first example we present will attempt to illustrate what a professional engineer might be called upon to do in a few minutes with no references. It is not the type of problem you, as a beginning student, would be expected to do because your have not yet gained the necessary experience.

Example problem 3.1   A civil engineer is asked to meet with a city council committee to discuss their needs with respect to the disposal of solid wastes (garbage or refuse). The community, a city of 12 000 persons, must begin supplying refuse collection and disposal for its citizens for the first time. In reviewing various alternatives for disposal, a sanitary landfill is suggested. One of the council members is concerned about how much land is going to be needed, so he asks the engineer how many acres will be required within the next 10 years.

Discussion   The engineer quickly estimates as follows:
The national average solid-waste production is 2.75 kg/(capita)(day). We can determine that each citizen will thus produce 1 000 kg of refuse per year by the following calculation:

(2.75 kg/day)(365 days/year) $\cong$ 1 000 kg/year

Experience indicates that refuse will probably be compacted to a density of 400 to 600 kg/m^3. On this basis, the per capita landfill volume will be 2 m^3 each year; and 1 acre filled 1 m deep will contain the collected refuse of 2 000 people for a year (1 acre = 4 047 m^2). Therefore, the requirement for 12 000 people will be 1 acre filled 6 m deep. However, knowledge of the geology of the particular area indicates that bedrock occurs at approximately 6 m below the ground surface. The completed landfill should therefore have an average depth of 4 m; consequently, 1.5 acres a year, or 15 acres in 10 years, will be required. The patterns of the recent past indicate that some growth in population and solid-waste generation should be expected. It is finally suggested that the city should plan to use about 20 acres in the next 10 years.

   This calculation took only minutes and required no computational device other than pencil and paper. The engineer's experience, rapid calculations, sound basic assumptions, and sensible rounding of figures were the main requirements. And a usable estimate, designed to neither mislead nor to sell a point of view, was provided. If this project proceeds to the actual development of a sanitary landfill, the civil engineer will then gather actual data, refine the calculations, and prepare estimates upon which one would risk a professional reputation.

Example prob. 3.2 is an illustration of a problem you might be assigned. Here you have the necessary experience to perform the estimation. Not counting the final written presentation, you should be able to do a similar problem in one-half to 1 hour.

Example problem 3.2   Suppose that your instructor assigns the following problem: Determine the number of pieces of lumber 5 cm $\times$ 10 cm $\times$ 2.40 m that can be sawn from the tree nearest to the southeast corner of the building in which you are now meeting. How would you proceed? See Fig. 3.5 for one student's response.

PROBLEM

ESTIMATE THE NUMBER OF 5cm × 10cm × 2.40m BOARDS THAT CAN BE SAWN FROM THE FIR TREE NEAR THE S.E. CORNER OF THE ENGINEERING BUILDING.

ASSUMPTIONS

1. THE TREE TRUNK IS CONICAL.
2. THE LIMBS CANNOT BE USED – TOO SMALL.
3. ALL PIECES THAT ARE TOO SMALL WILL BE DISCARDED – NO PARTIALS OR PARTICLE BOARDS.
4. THE TREE WILL BE CUT 0.3m ABOVE THE GROUND.

COLLECTED DATA

1. I AM 180cm TALL AND MY SHADOW WAS 135cm.
2. THE TREES SHADOW WAS 14m.
3. THE BASE OF THE TREE HAS A CIRCUMFERENCE OF 120cm.

SOLUTION

HEIGHT OF TREE $\left(\frac{180}{135}\right)$ 14 m = 18.67m – APPROX ~ 19m

DIAMETER OF TREE AT GROUND $\frac{120}{\pi}$ = 38.2cm

APPROX ~ 38cm

DIAMETER REDUCTION = $\frac{38cm}{19m}$ = 2.0 $\frac{cm}{m}$

FIRST SECTION (0.3m – 2.7m)

EFFECTIVE CROSS SECTION @ 2.7m MEASURING FROM GROUND = 38cm – (2.7 × 2.0) = 32.6 cm ~ 33 cm

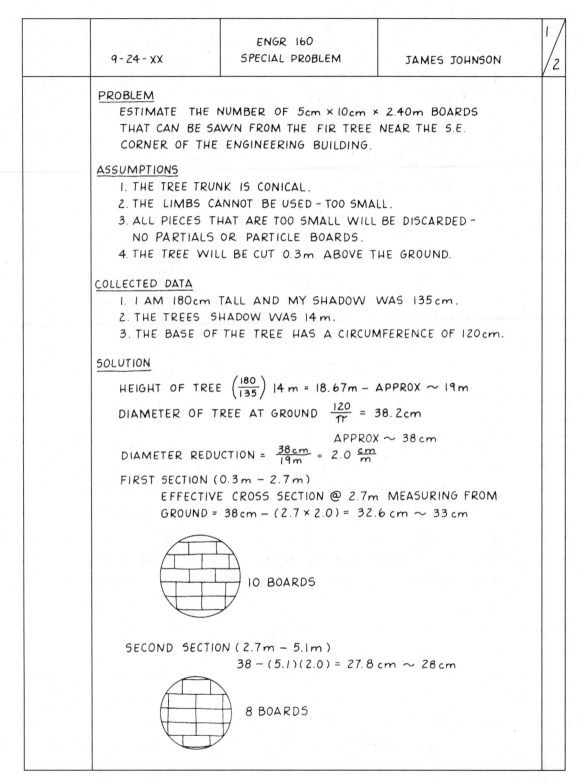

10 BOARDS

SECOND SECTION (2.7m – 5.1m)

38 – (5.1)(2.0) = 27.8 cm ~ 28cm

8 BOARDS

Figure 3.5
Student presentation for
Example prob. 3.2.

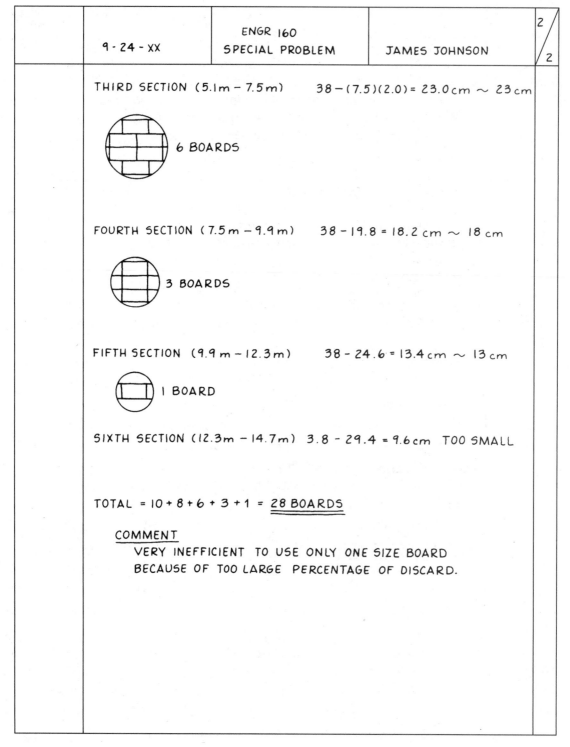

THIRD SECTION (5.1m – 7.5m)   38 – (7.5)(2.0) = 23.0 cm ~ 23 cm

6 BOARDS

FOURTH SECTION (7.5m – 9.9m)   38 – 19.8 = 18.2 cm ~ 18 cm

3 BOARDS

FIFTH SECTION (9.9 m – 12.3 m)   38 – 24.6 = 13.4 cm ~ 13 cm

1 BOARD

SIXTH SECTION (12.3m – 14.7m)   3.8 – 29.4 = 9.6 cm   TOO SMALL

TOTAL = 10 + 8 + 6 + 3 + 1 = <u>28 BOARDS</u>

<u>COMMENT</u>
VERY INEFFICIENT TO USE ONLY ONE SIZE BOARD
BECAUSE OF TOO LARGE PERCENTAGE OF DISCARD.

Figure 3.5 (cont.)

Discussion   The assumptions that Jim made seem to be reasonable. Although he did not allow for the width of the saw cut nor for the thickness of the bark, the omissions may very well be consistent with the degree of accuracy involved in his calculations. Most Boy and Girl Scouts have learned to measure heights by the method he used; and all freshman engineering students should be familiar with similar triangles. After determining the height and diameter of the tree, Jim applied a graphical technique for determining how many boards could be cut from each 2.4 m section of the tree. He correctly used the upper (smaller) diameter of the section. His task was then reduced to a simple counting of the boards.

We have mentioned previously that both time allotted and physical circumstances (reference material at hand, access to a phone, and so on are major factors that influence the type of estimate you can produce. Example probs. 3.3 and 3.4 will show reasonable estimates of the same quantity (paper used) under two different sets of circumstances. In the first case, the student had about 10 minutes to do the estimate under exam conditions; that is, the student could not leave the classroom seat and had no reference material. The second case resulted from a homework assignment where 1 to 2 hours were available and the student could obtain data needed for the estimate.

Example problem 3.3   Estimate the amount of paper used by the students of this college for homework, quizzes, and examinations during this academic year. Express your answer in kilograms. You may use no reference materials.

Discussion   The problem requires that data be known or assumed. The calculation is simple and straightforward once the data are established (see Fig. 3.6). Because much data must be assumed without verification, the estimate cannot be expected to be very precise, so only one, or at most two, significant figures should be reported. Two significant figures represent a maximum error range of about 10 percent. For example, 4 200 represents numbers between 4 150 and 4 250, an error of 100 (4 200 ± 50) in 4 200 or about 2.4 percent.

Example problem 3.4   Estimate the amount of paper used by the students of this college for homework, quizzes, and examinations during this academic year. Express your answer in kilograms. Plan to use 1 to 2 hours as a homework assignment.

Discussion   The results of this estimation are found in Fig. 3.7. Many of the assumptions that had to be made in Example prob. 3.3 are no longer necessary because time is available to obtain information. The small sample in the survey still produces some uncertainty. In each case, your reader should be reasonably convinced by your techniques; for example, anyone would be willing to accept student population obtained from the registrar.

	3-16-XX	ENGR 160 PROBLEM 4.11	SARA KENILWORTH	1/1

PROBLEM 4.11

ESTIMATE THE MASS (kg) OF PAPER USED BY STUDENTS IN THIS COLLEGE DURING THIS ACADEMIC YEAR FOR HOMEWORK, QUIZZES, AND EXAMS.

ASSUMPTIONS

1. COLLEGE HAS 4200 STUDENTS.
2. 2 SHEETS ARE USED PER CREDIT HOUR PER WEEK.
3. 32 WEEKS PER ACADEMIC YEAR.
4. TABLET OF 50 SHEETS OF PAPER HAS A MASS OF ABOUT ½ lbm.
5. STUDENTS AVERAGE 15 CREDITS PER TERM.

CALCULATIONS

$$\text{MASS OF SHEET} = \left(\frac{0.50\ \text{lbm}}{50\ \text{SHEETS}}\right)\left(\frac{454\ \text{g}}{\text{lbm}}\right) = 4.54\ \text{g/SHEET}$$

$$\text{MASS OF PAPER} = (4200\ \text{STUDENTS})\left(\frac{15\ \text{CREDITS}}{\text{STUDENT}}\right)\left(\frac{2\ \text{SHEETS}}{(\text{CREDIT})(\text{WEEK})}\right)\left(\frac{32\ \text{WEEKS}}{\text{YEAR}}\right)\left(\frac{4.54\ \text{g}}{\text{SHEET}}\right)\left(\frac{1\ \text{Kg}}{1000\ \text{g}}\right)$$

$$= 18305.28\ \text{kg/YEAR}$$

$$\cong 1.8 \times 10^4\ \text{kg/YEAR}$$

Figure 3.6
Student presentation for Example prob. 3.3.

PROBLEM 4.11

ESTIMATE THE MASS (kg) OF PAPER USED BY STUDENTS IN THIS
COLLEGE DURING THIS ACADEMIC YEAR FOR HOMEWORK,
QUIZZES, AND EXAMS.

ASSUMPTIONS

1. THIS COLLEGE REFERS TO THE COLLEGE OF ENGINEERING, STATE UNIVERSITY
2. ONLY UNDERGRADUATE STUDENTS WILL BE CONSIDERED.
3. THE COMPUTATION IS FOR THE 1983-84 ACADEMIC YEAR.
4. ENGINEERING-PROBLEMS PAPER IS THE STANDARD FOR THE COLLEGE.
5. THE RESULTS OF THE SURVEY BELOW REASONABLY REPRESENT
   THE ENGINEERING STUDENTS AS A WHOLE.

COLLECTED DATA

1. ACCORDING TO THE UNIVERSITY REGISTRAR, 4 256 ENGINEERING
   UNDERGRADUATES WERE REGISTERED FOR FALL SEMESTER AND
   4 028 FOR SPRING SEMESTER.
2. THE GENERAL CATALOG SHOWS 15 WEEKS OF CLASSES EACH
   SEMESTER PLUS FINAL EXAM WEEK EACH TERM.
3. A SURVEY OF 8 ENGINEERING STUDENTS, 2 FROM EACH CLASS,
   FRESHMAN THROUGH SENIOR, PROVIDED THE FOLLOWING DATA:

STUDENTS	SHEETS/WEEK	SHEETS/EXAM WEEK	COMMENTS
1	35	15	
2	18	8	PART TIME
3	45	12	
4	52	20	20 CREDIT HOURS
5	28	10	
6	25	14	
7	38	15	
8	42	18	
AVE	35.4	14	

4. A REAM (500 SHEETS) OF ENGINEERING - PROBLEMS PAPER HAS A
   MASS OF 3.75 lbm AS DETERMINED WITH A POSTAL SCALE.

THEORY

MASS PER SHEET = (MASS PER REAM) / (500 SHEETS PER REAM)

MASS OF PAPER = MASS FOR FALL SEMESTER + MASS FOR SPRING SEMESTER.

WHERE

MASS FOR FALL = (NUMBER OF STUDENTS) [(SHEETS/WEEK) (CLASS WEEKS) +
(SHEETS/EXAM WEEK)] (MASS/SHEET)

MASS FOR SPRING = (NUMBER OF STUDENTS) [(SHEETS/WEEK) (CLASS WEEKS) +
(SHEETS/EXAM WEEK)] (MASS/SHEET)

Figure 3.7
Student presentation for
Example prob. 3.4.

	3-16-XX	ENGR 160 PROBLEM 4.11	SARA KENILWORTH	2/2

CALCULATIONS

MASS PER SHEET = [(3.75 lbm PER REAM)/(500 SHEETS)](0.453 59 kg/lbm)
= 0.003 402 kg/SHEET

MASS FOR FALL = (4 256 STUDENTS)[(35.4 SHEETS/WEEK)(15 WEEKS) + 14 SHEETS](0.003 402 kg/SHEET)
= 7 891 kg

MASS FOR SPRING = (4 028 STUDENTS)[(35.4 SHEETS/WEEK)(15 WEEKS) + 14 SHEETS](0.003 402 kg/SHEET)
= 7 468 kg

TOTAL MASS OF PAPER = 7 891 kg + 7 468 kg
= 15 359 kg

ESTIMATE OF TOTAL MASS = 1.5×10^4 kg

Figure 3.7 (cont.)

Another type of estimate that engineers are called upon to make is one where a choice is involved. Estimates are prepared of the various alternatives available, and a decision as to which alternative to follow is then made; perhaps the one that is least expensive is chosen. Example prob. 3.5 is designed to show estimating for decision purposes. This problem requires experience that a student might have.

Example problem 3.5 Approximate the amount of gasoline that will be used by the students of Iowa State University during the next Christmas–New Year recess for the purpose of visiting their homes and returning. Provide the answer in gallons. What will be the total cost of this gasoline?

Discussion Figure 3.8 is the result of the approximation. A number of assumptions were made in order to obtain a solution (without writeup) in less than 30 minutes. Note that the writeup was prepared on a word processor.

PROBLEM 3.5

Estimate the amount of gasoline that will be used by Iowa State University in traveling home and returning during the upcoming Christmas-New Year's holiday break. Also estimate the cost of the gasoline.

ASSUMPTIONS

1. Only automobile usage is considered.
2. Ten percent of the students fly home. Only the trip to the airport will be considered.
3. Ten percent of the students remain on campus.
4. An average of two students travel together in the same automobile.
5. The average automobile used for the trip gets 20 miles to the gallon.
6. Travel distance will be expressed in terms of median round trip distances of 100, 300, 500, 700, 900 miles. For longer distances, students are assumed to fly.

COLLECTED DATA

1. Iowa State has 24 500 students.   (FROM STUDENT PROFILES BROCHURE)
   Iowa residents 65%
   Other U.S.   25%
   International 10%
2. Cost of gasoline is $1.08/gal
3. Distance to airport is 45 miles

Figure 3.8
Student presentation, produced on a word processor, for Example prob. 3.5.

Travel distance estimate:

round trip distance	percent of students	number of students
0-200	35	8 575
200-400	30	7 350
400-600	5	1 225
600-800	5	1 225
800-1000	5	1 225
Flying	10	2 450
Not traveling	10	2 450
	100	24 500

Mileage total:

0-200 mi   [(8 575 students)/(2 students/car)][100 mi/car]   =   428 750 mi

200-400 mi   [7 350 /2][300]   = 1 102 500 mi

400-600 mi   [1 225/2][500]   =   306 250 mi

600-800 mi   [1 225/2][700]   =   428 750 mi

800-1000 mi   [1 225/2][900]   =   551 250 mi

Flying   [2 450/2][90]   =   110 250 mi

                                                                 2 927 750 mi

gasoline amount:

(2 927 750 mi)/(20 mi/gal) = 150 000 gal

gasoline cost:

(150 000 gal)($1.08/gal) = $160 000

70

Physical measurement	Random error (in a	**Key Terms and**
Significant digits	measurement)	**Concepts**

Physical measurement
Significant digits
Scientific notation
Accuracy
Precision
Systematic error (in a
   measurement)

Random error (in a
   measurement)
Engineering estimates
Engineering approximations

3.1 How many significant digits are contained in each of the following quantities?

    (a) 4 930   (f) 6.220×10^2
    (b) 4.930   (g) 9.009
    (c) 0.049 3   (h) 0.000 3
    (d) 206.0   (i) 0.320 0
    (e) 200   (j) 40 620

3.2 How many significant digits are contained in each of the following quantities?

    (a) 322   (f) 0.0030
    (b) 3.22   (g) 3 600 s/h
    (c) 0.032 2   (h) 5 280 ft/mi
    (d) 0.032 20   (i) 2.006×10^4
    (e) $3.220\ 3 \times 10^5$ (j) 0.090 50

3.3 Perform the following computations and report the answers with the proper number of significant digits.

    (a) (2.05)(360)
    (b) 26.35/14
    (c) [(4.91)(32.2)]/12.03
    (d) $[(14.2 \times 10^3)(7.3 \times 10^{-1})(1\ 021)]/[(16)(21.0 \times 10^{-3})]$
    (e) $(4.597\ 6 \times 10^2) + (4.721 \times 10^1)$
    (f) 0.009 024 + 0.065 320
    (g) (0.127 60 − 0.073 2)/14.06
    (h) 163/(0.021 68 − 0.021 66)

3.4 Perform the following computations. Record results with the proper number of significant digits.

    (a) 648.2/14.0
    (b) 436.2 mi/1.06h
    (c) $(4.010 \times 10^3 \text{ s})/(3\ 600 \text{ s/h})$
    (d) ($95.99)(0.15)
    (e) $[(4.83 \times 10^3) - (14.63 \times 10^2)]/136.2$

3.5 Express the results of the following computations with the proper number of significant digits.

    (a) $Q = 4.0P^2 - 35\ P - 23$ for $P = 10.4$
    (b) $(46.13)^2/4$
    (c) $Y = 4.33 \sin x - 12.0 \sin x \cos x$ for $x = 1.2$ radians
    (d) ($36 ticket)(23 850 tickets)
    (e) $[(3.0004)/(8.660\ 0 \times 10^{-3})] - 345.6$

3.6 (a) The accuracy of a multimeter is given as ± 0.8 percent of full scale. If the full scale reading is 130 volts, what is the range of values associated with a full scale reading?

(*b*) A thermometer is found to yield readings that are consistently 2.5 percent high. If a reading of 82.4°F is taken, what is the correct temperature?

3.7 (*a*) Compute the percent error for the following cases.

	Case A	Case B
Measured	1 203.7	17.4
True	1 204.8	16.3

(*b*) From the results in part a, discuss the precision of the measurements in the two cases.

3.8 Estimate the total volume in cubic meters and mass in kilograms of the concrete and asphalt paving in the largest parking lot on your campus.

3.9 Estimate the volume (capacity) in cubic meters of a water tower located on your campus or nearby (as specified by your instructor).

3.10 How much paint (primer plus two coats) is needed to paint a water tower on your campus or nearby (as specified by your instructor)? Consider only the tank portion, not the supporting structure, if any. Express result in gallons.

3.11 Compute the surface area (in square meters) of external glass in your classroom building or any other structure designated by your instructor. What is the mass of this glass in kilograms?

3.12 Suppose your school's basketball team plans to play in a 3-day tournament in Kansas City. Your job is to estimate the total cost of transporting the team, food and housing for the team, and so on. Consider the essential personnel (team, coaches, and managers) only.

3.13 Your school has plans to require all entering engineering students to purchase their own microcomputers. Estimate the total cost to students for the next fall term based on a system specification worked out by the class or provided by the instructor.

3.14 Based on your local electric energy rates, estimate the cost of lighting for your classroom for this academic year.

3.15 By your personal observation, estimate the total number of computer terminals available for general student use on your campus. Determine the maximum number of students that could be served for 2 hours each week, assuming the terminals are available 24 hours per day, 7 days per week. How does this compare with the number of students on campus?

3.16 Estimate the number of audio tapes and CDs owned by a typical engineering student at your school. Include a well-documented survey in your solutions.

3.17 Each year rain and/or snow falls on your campus. For a specified parking lot or rooftop, estimate the total volume (in cubic meters) and mass (in kilograms) of water that must be carried away during the entire year.

3.18 By utilizing a reasonable survey technique, compute the amount of money annually spent by all engineering students enrolled at your school for long-distance telephone calls.

3.19 If your campus has a body of water (lake, pond, or fountain), determine the volume of water contained in it (in cubic feet) and the surface area (in square feet). Do the estimate without getting wet or getting in trouble with campus security.

3.20 Estimate the mass of all textbooks you will purchase to complete your first engineering degree. If you have not selected a major, use one that is a likely candidate.

3.21 What is the total floor area (in square feet) and volume (in cubic feet) of the classrooms in the building where this course meets?

3.22 Assuming that the interior of the classroom where this class meets is painted, estimate the amount of paint necessary to refurbish it and make a significant change in wall color. (Alternative: Do the estimate for another interior space designated by your instructor.)

3.23 Following an accident where a student was cut by broken glass in a door, your school has decided to replace all door glass with plastic (Plexiglas or some other brand) in the building where this class meets. Considering interior and exterior doors, estimate the amount (in square feet) of material needed and its approximate cost. Exclude labor costs in this estimate.

3.24 For a building (or portion of a building) on your campus that is carpeted (the student union building, perhaps), estimate the amount (in square yards) of material reqiured to recarpet the space. What is the expected cost of the materials?

Problems 3.25 through 3.28 require a significant amount of data to be collected prior to performing the estimates. You should allow at least 2 hours (excluding writeup) to complete each problem.

3.25 Estimate the amount of each of the ingredients required to make the concrete used in all the designated interstate highways in your state.

3.26 Estimate the amount of water used in a 1-year period by a family of four who own their own house. Determine the cost from local utility rates.

3.27 Estimate the number of family dwellings that could be supplied by the electricity generated by all the nuclear power plants in the United States.

3.28 Estimate the distance (in kilometers) that you walk (and run) during a typical week while attending classes at your school. Include all activities each day.

Dimensions, Units, and Conversions

Years ago, when countries were more isolated from one another, individual governments tended to develop and use their own set of measures. As the rapid increase in global communication and travel brought countries closer together and the world advanced in technology, the need for a universal system of measurement became abundantly clear. There was such a growth of information but diversity of reporting among nations that a standard set of dimensions, units, and measurements was vital if this wealth of knowledge was to be of benefit at all. This chapter deals with the difference between dimensions and units and at the same time explains how there can be an orderly transition from many systems of units to one system—that is, an international standard.

The standard currently accepted in most industrial nations (it is optional in the United States) is the international metric system, or Systeme International d'Unites, abbreviated SI. The SI units are a modification and refinement of an earlier version of the metric system (MKS) that designated the meter, kilogram, and second as fundamental units.

France was the first country, in 1840, to officially legislate adoption of the metric system and decree that its use be mandatory.

The United States almost adopted the metric system 150 years ago. In fact, the metric system was made legal in the United States in 1866, but its use was not made compulsory. In spite of many attempts since that time, full conversion to the metric system has not yet been realized in the United States, but significant steps in that direction are continuously underway.

Engineers are constantly concerned with the measurements of fundamental physical quantities such as length, time, temperature, force, and so on. In order to specify a physical quantity fully, it is not sufficient to indicate merely a numerical value. The magnitude of physical quantities can be understood only when they are compared with predetermined reference amounts, called *units*. Any measurement is, in effect, a comparison of how many

Look for the **km/h** tab below the maximum speed limit sign, indicating that this is the new speed in metric.

100 km/h This speed limit will likely be the most common on freeways. On most rural two-lane roadways, **80 km/h** will be typical.

50 km/h A **50 km/h** speed limit will apply in most cities.

Actual speed limits will be established in accordance with local regulations.

Metric Commission Canada   Commission du système métrique Canada

Surveillez l'indication de l'unité de vitesse **km/h**; ce symbole signifie que va vitesse est mesurée selon le système métrique.

100 km/h Sur les autoroutes, la vitesse maximale la plus courante sera de **100 km/h** tandis que sur les routes à grande circulation, elle sera de **80 km/h**.

50 km/h Dans la plupart des grands centres, la vitesse maximale sera de **50 km/h**.

Les vitesses maximales en vigueur dans votre société seront établies selon les règlements municipaux.

Commission du système métrique Canada   Metric Commission Canada

Figure 4.1
Highway signs in Canada.
(Metric Commission of Canada.)

(the numerical value V) units U are contained within the physical quantity Q. If we consider length the physical quantity L, and 20.0 the numerical value V, with meters as the designated units U, then $L = 20.0$ m can be represented in a general fashion by the expression

$$Q = VU$$

For this relationship to be valid, the exact reproduction of a unit must be theoretically possible at any time. Therefore standards must be established. These standards are a set of fundamental unit quantities kept under normalized conditions in order to preserve their values as accurately as possible. We shall speak more about standards and their importance later.

4.3

Dimensions Dimensions are used to describe physical quantities. An important element to remember is that dimensions are independent of units. As outlined above, the physical quantity length can be represented by the dimension L, for which there are a large number of possibilities available when selecting a unit. For example, in ancient Egypt, the cubit was related to the length of the arm from the tip of the middle finger to the elbow. Measurements were thus a function of physical stature, with variation from one individual to another. Much later, in Britain, the inch was specified as the distance covered by three barley corns, round and dry, laid end to end.

Today we require more precision. For example, the meter is defined in terms of the distance traveled by light in a vacuum in a specified amount of time. We can draw two important points from this discussion: (1) Physical quantities can be accurately measured, and (2) each of these units (cubit, inch, and meter), although distinctly different, has in common the quality of being a length and not an area or a volume.

A technique used to distinguish between units and dimensions is to call all quantities of length simply L. In this way, each new physical quantity gives rise to a new dimension, such as T for time, F for force, M for mass, and so on. (Note that there are as many dimensions as there are kinds of physical quantities.)

Moreover, dimensions can be divided into two areas—fundamental and derived. A fundamental dimension is a dimension that can be conveniently and usefully manipulated when expressing all physical quantities of a particular field. Derived dimensions are a combination of fundamental dimensions. Velocity, for example, could be defined as fundamental dimension V, but it is more customary as well as more convenient to consider velocity as a combination of fundamental dimensions, so that it becomes a derived dimension, $V = (L)(T)^{-1}$. L and T are fundamental dimensions, and V is a derived dimension because it is made up of two fundamental dimensions (L,T).

It is advantageous to use as few fundamental dimensions as possible, but the selection of what is to be fundamental and what is to be derived is not fixed. In actuality, any dimension can be selected as a fundamental dimension in a particular field of engineering or science; and for reasons of convenience, it may be a derived dimension in another field.

A *dimensional system* can be defined as the smallest number of fundamental dimensions which will form a consistent and complete set for a field of science. For example, three fundamental dimensions are necessary to form a complete mechanical dimensional system. Depending on the discipline these dimensions may be specified as either length (L), time (T), and mass (M) or length (L), time (T), and force (F). If temperature is important to the application, a fourth dimension must be added.

The *absolute system* (so called because dimensions used are not affected by gravity) has as its fundamental dimensions L, T, and M. An advantage of this system is that comparisons of masses at various locations can be made with an ordinary balance, because the local acceleration of gravity has no influence upon the results.

Table 4.1 Two basic dimensional systems

Quantity	Absolute	Gravitational
Length	L	L
Time	T	T
Mass	M	$FL^{-1}T^2$
Force	MLT^{-2}	F
Velocity	LT^{-1}	LT^{-1}
Pressure	$ML^{-1}T^{-2}$	FL^{-2}
Momentum	MLT^{-1}	FT
Energy	ML^2T^{-2}	FL
Power	ML^2T^{-3}	FLT^{-1}
Torque	ML^2T^{-2}	FL

The *gravitational system* has as its fundamental dimensions L, T, and F. It is widely used in many engineering branches because it simplifies computations when weight is a fundamental quantity in the computations. Table 4.1 illustrates two of the more basic dimensional systems; however, a number of other dimensional systems are commonly used for heat, electromagnetism, electrical dimensions, and so forth.

4.4

Units

Once a consistent dimensional system has been selected, one must select a unit system by choosing a specific unit for each fundamental dimension. The problem one encounters when working with units is that there can be a large number of unit systems to choose from for each complete dimension system, as we have already suggested. It is obviously desirable to limit the number of systems and combinations of systems. The SI previously alluded to is intended to serve as an international standard that will provide worldwide consistency.

There are two fundamental systems of units commonly used in mechanics today. One system used in almost every industrial country of the world is called the *metric system*. It is a decimal-absolute system based on the meter, kilogram, and second (MKS) as the units of length, mass, and time, respectively. The United States has used the other system, normally referred to as the British gravitational system. It is based on the foot, pound-force, and second.

Numerous international conferences on weights and measures over the past 40 years have gradually modified the MKS system to the point that all countries previously using various forms of the metric system are beginning to standardize. The Systeme International d'Unites (SI) is now considered the new international system of units. The United States has adopted the system, but full use will be preceded by a long and expensive period of change. During this transition period, engineers will have to not only be familiar with SI but also other systems and the necessary conversion process between or among systems.

SI Units and Symbols

The International System of Units (SI), developed and maintained by the General Conference on Weights and Measures (Conference Generale des Poids et Mesures, CGPM), is intended as a basis for worldwide standardization of measurements. The name and abbreviation were set forth in 1960. SI at the present time is a complete system that is being universally adopted.

This new international system is divided into three classes of units:

1. Base units
2. Supplementary units
3. Derived units

There are seven base units in the SI. The units (except the kilogram) are defined in such a way that they can be reproduced anywhere in the world.

Table 4.2 lists each base unit along with its name and proper symbol.

In the following list, each of the base units is defined as established at the international CGPM:

1. Length: The meter is a length equal to the distance traveled by light in a vacuum during 1/299 792 458 s. The meter was defined by the CGPM that met in 1983.
2. Time: The second is the duration of 9 192 631 770 periods of radiation corresponding to the transition between the two hyperfine levels of the ground state of the cesium-133 atom. The second was adopted by the thirteenth CGPM in 1967.
3. Mass: The standard for the unit of mass, the kilogram, is a cylinder of platinum-iridium alloy kept by the International Bureau of Weights and Measures in France. A duplicate copy is maintained in

Table 4.2 Base units

Quantity	Name	Symbol
Length	meter	m
Mass	kilogram	kg
Time	second	s
Electric current	ampere	A
Thermodynamic temp.	kelvin	K
Amount of substance	mole	mol
Luminous intensity	candela	cd

the United States. The unit of mass was adopted by the First and Third CGPMs in 1889 and 1901. It is the only base unit nonreproducible in a properly equipped lab.

4. Electric current: The ampere is a constant current which, if maintained in two straight parallel conductors of infinite length and of negligible circular cross sections and placed one meter apart in volume, would produce between these conductors a force equal to 2×10^{-7} newton per meter of length. The ampere was adopted by the Ninth CGPM in 1948.

5. Temperature: The kelvin, a unit of thermodynamic temperature, is the fraction 1/273.16 of the thermodynamic temperature of the triple point of water. The kelvin was adopted by the Thirteenth CGPM in 1967.

6. Amount of substance: The mole is the amount of substance of a system that contains as many elementary entities as there are atoms in 0.012 kilogram of carbon-12. The mole was defined by the Fourteenth CGPM in 1971.

7. Luminous intensity: The base unit candela is the luminous intensity in a given direction of a source that emits monochromatic radiation of frequency 540×10^{12} hertz and has a radiant intensity in that direction of 1/683 watts per steradian.

The units listed in Tab. 4.3 are called *supplemental units* and may be regarded as either base units or as derived units.

The unit for a plane angle is the radian, a unit that is used frequently in engineering. The steradian is not as commonly used. These units can be defined in the following way:

1. Plane angle: The radian is the plane angle between two radii of a circle that cut off on the circumference of an arc equal in length to the radius.

2. Solid angle: The steradian is the solid angle which, having its vertex in the center of a sphere, cuts off an area of the sphere equal to that of a square with sides of length equal to the radius of the sphere.

As indicated earlier, derived units are formed by combining base, supplementary, or other derived units. Symbols for them are carefully selected to avoid confusion. Those which have special names and symbols, as interpreted for the United States by the National Bureau of Standards, are listed in Tab. 4.4 together with their definitions in terms of base units.

Table 4.3 Supplemental units

Quantity	Name	Symbol
Plane angle	radian	rad
Solid angle	steradian	sr

Table 4.4 Derived units

Quantity	SI unit symbol	Name	Base units
Frequency	Hz	hertz	s^{-1}
Force	N	newton	$kg \cdot m \cdot s^{-2}$
Pressure stress	Pa	pascal	$kg \cdot m^{-1} \cdot s^{-2}$
Energy or work	J	joule	$kg \cdot m^2 \cdot s^{-2}$
Quantity of heat	J	joule	$kg \cdot m^2 \cdot s^{-2}$
Power radiant flux	W	watt	$kg \cdot m^2 \cdot s^{-3}$
Electric charge	C	coulomb	$A \cdot s$
Electric potential	V	volt	$kg \cdot m^2 \cdot s^{-3} \cdot A^{-1}$
Potential difference	V	volt	$kg \cdot m^2 \cdot s^{-3} \cdot A^{-1}$
Electromotive force	V	volt	$kg \cdot m^2 \cdot s^{-3} \cdot A^{-1}$
Capacitance	F	farad	$A^2 \cdot s^4 \cdot kg^{-1} \cdot m^{-2}$
Electric resistance	Ω	ohm	$kg \cdot m^2 \cdot s^{-3} \cdot A^{-2}$
Conductance	S	siemens	$kg^{-1} \cdot m^{-2} \cdot s^3 \cdot A^2$
Magnetic flux	Wb	weber	$kg^{-1} \cdot m \cdot s^{-2} \cdot A^{-1}$
Magnetic flux density	T	tesla	$kg \cdot s^2 \cdot A^{-1}$
Inductance	H	henry	$kg \cdot m^2 \cdot s^{-2} \cdot A^{-2}$
Luminous flux	lm	lumen	$cd \cdot sn$
Illuminance	lx	lux	$cd \cdot sn \cdot m^{-2}$
Celsius temperature*	°C	degree Celsius	K
Activity (radionuclides)	Bq	becqueret	s^{-1}
Absorbed dose	Gy	gray	$m^2 \cdot s^{-2}$
Dose equivalent	S	sievert	$m^2 \cdot s^{-2}$

*The thermodynamic temperature (T_k) expressed in kelvins is related to Celsius temperature ($t_{°C}$) expressed in degrees Celsius by the equation $t_{°C} = T_k - 273.15$.

At first glance, Fig. 4.2 may appear complex, even confusing; however, a considerable amount of information is presented in this concise flowchart. To get the point of it quickly, be aware that the solid lines denote multiplication and the broken lines indicate division. The arrows pointing toward the units (circled) are significant and arrows going away have no meaning for that particular unit. Consider the pascal, as an example: Two arrows point toward the circle—one solid and one broken. This means that the unit pascal is formed from the newton and meter squared, or N/m^2.

Other derived units, such as those included in Tab. 4.5, have no special names but are combinations of base units and units with special names.

Being a decimal system, the SI is convenient to use because by simply affixing a prefix to the base, a quantity can be increased or decreased by factors of 10 and the numerical quantity can be kept within manageable limits. Table 4.6 lists the multiplication factors with their prefix names and symbols.

The proper selection of prefixes will also help eliminate non-significant zeros and leading zeros in decimal fractions. One rule to follow is that the numerical value of any measurement should be recorded as a number between 0.1 and 1 000. This rule is suggested because it is easier to make realistic judgments when working with numbers between 0.1 and 1 000. For example, suppose that you are asked the distance to a nearby town. It would

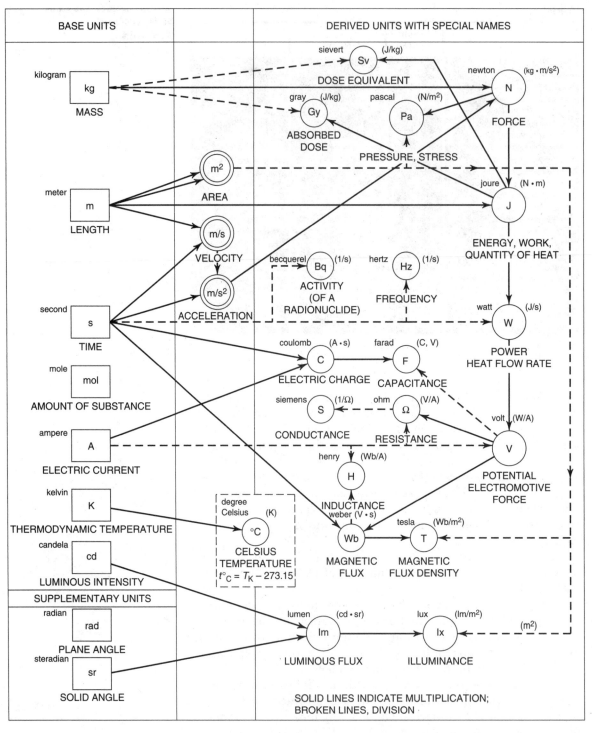

Figure 4.2
Graphical illustration of how certain SI units are derived in a coherent fashion from base and supplementary units. *(National Bureau of Standards.)*

Table 4.5 Common derived units

Quantity	Units	Quantity	Units
Acceleration	$m \cdot s^{-2}$	Molar entropy	$J \cdot mol^{-1} \cdot K^{-1}$
Angular acceleration	$rad \cdot s^{-2}$	Molar heat capacity	$J \cdot mol^{-1} \cdot K^{-1}$
Angular velocity	$rad \cdot s^{-1}$	Moment of force	$N \cdot m$
Area	m^2	Permeability	$H \cdot m^{-1}$
Concentration	$mol \cdot m^{-3}$	Permittivity	$F \cdot m^{-1}$
Current density	$A \cdot m^{-2}$	Radiance	$W \cdot m^{-2} \cdot sr^{-1}$
Density, mass	$kg \cdot m^{-3}$	Radiant intensity	$W \cdot sr^{-1}$
Electric charge density	$C \cdot m^{-3}$	Specific heat capacity	$J \cdot kg^{-1} \cdot K^{-1}$
Electric field strength	$V \cdot m^{-1}$	Specific energy	$J \cdot kg^{-1}$
Electric flux density	$C \cdot m^{-2}$	Specific entropy	$J \cdot kg^{-1} \cdot K^{-1}$
Energy density	$J \cdot m^{-3}$	Specific volume	$m^3 \cdot kg^{-1}$
Entropy	$J \cdot K^{-1}$	Surface tension	$N \cdot m^{-1}$
Heat capacity	$J \cdot K^{-1}$	Thermal conductivity	$W \cdot m^{-1} \cdot K^{-1}$
Heat flux density	$W \cdot m^{-2}$	Velocity	$m \cdot s^{-1}$
Irradiance	$W \cdot m^{-2}$	Viscosity, dynamic	$Pa \cdot s$
Luminance	$cd \cdot m^{-2}$	Viscosity, kinematic	$m^2 \cdot s^{-1}$
Magnetic field strength	$A \cdot m^{-1}$	Volume	m^3
Molar energy	$J \cdot mol^{-1}$	Wavelength	m

be more understandable to respond in kilometers than meters. That is, it is easier to visualize 10 km than 10 000 m.

Moreover, the use of certain prefixes is preferred over that of others. Those representing powers of 1 000, such as kilo-, mega-, milli-, and micro-, will reduce the number you must remember. These preferred prefixes should be used, with the following three exceptions which are still common because of convention:

1. When expressing area and volume, the prefixes hecto-, deka-, deci-, and centi- may be used, for example, cubic centimeter.

2. When discussing different values of the same quantity or expressing them in a table, calculations are simpler to perform when you use the same unit multiple throughout.

Table 4.6 Decimal multiples

Multiplier	Prefix name	Symbol
10^{18}	exa	E
10^{15}	peta	P
10^{12}	tera	T
10^{9}	giga	G
10^{6}	*mega	M
10^{3}	*kilo	k
10^{2}	hecto	h
10^{1}	deka	da
10^{-1}	deci	d
10^{-2}	centi	c
10^{-3}	*milli	m
10^{-6}	*micro	μ
10^{-9}	nano	n
10^{-12}	pico	p
10^{-15}	femto	f
10^{-18}	atto	a

*Most often used.

3. Sometimes a particular multiple is recommended as a consistent unit even though its use violates the 0.1 to 1 000 rule. For example, many companies use the millimeter for linear dimensions even when the values lie far outside this suggested range. The cubic decimeter (commonly called liter) is also used.

Recalling the discussion of significant figures, we see that SI prefix notations can be used to a definite advantage.

Consider the previous example of 10 km. When giving an estimate of distance to the nearest town, there is certainly an implied approximation in the use of a round number. Suppose that we were talking about a 10 000 m Olympic track and field event. The accuracy of such a distance must certainly be greater than something between 5 000 and 15 000 m. This example is intended to illustrate the significance of the four zeros. If all four zeros are in fact significant, then the race is accurate within 1 m (9 999.5 to 10 000.5). If only three zeros are significant, then the race is accurate to within 10 m

(9 995 to 10 005).

There are two logical and acceptable methods available of eliminating confusion concerning zeros:

1. Use proper prefixes to denote intended significance.

Distance	Precision
10.000 km	9 999.5 to 10 000.5 m
10.00 km	9 995 to 10 005 m
10.0 km	9 950 to 10 050 m
10 km	5 000 to 15 000 m

2. Use scientific notation to indicate significance.

Distance	Precision
10.000×10^3 m	9 999.5 to 10 000.5 m
10.00×10^3 m	9 995 to 10 005 m
10.0×10^3 m	9 950 to 10 050 m
10×10^3 m	5 000 to 15 000 m

Selection of a proper prefix is customarily the logical way to handle problems of significant figures; however, there are conventions that do not lend themselves to the prefix notation. An example would be temperature in degrees Celsius; that is, $4.00(10^3)°C$ is the conventional way to handle it, not 4.00 k°C.

4.6

Rules for Using SI Units

Along with the adoption of SI comes the responsibility to thoroughly understand and properly apply the new system. Obsolete practices involving both English and metric units are widespread. This section provides rules that should be followed when working with SI units.

1. Periods are never used after symbols unless the symbol is at the end of a sentence (that is, SI unit symbols are not abbreviations).

2. Unit symbols are written in lowercase letters unless the symbol derives from a proper name, in which case the first letter is capitalized.

Lowercase	Uppercase
m, kg, s, mol, cd	A, K, Hz, Pa, C

3. Symbols rather than self-styled abbreviations should always be used to represent units.

Correct	Not correct
A	amp
s	sec

4. An s is never added to the symbol to denote plural.

5. A space is always left between the numerical value and the unit symbol.

Correct	Not correct
43.7 km	43.7km
0.25 Pa	0.25Pa

Exception: No space should be left between numerical values and the symbols for degree, minute, and second of angles and for degree Celsius.

6. There should be no space between the prefix and the unit symbols.

Correct	Not correct
mm, MΩ	k m, μ F

7. When writing unit names, all letters are lowercase except at the beginning of a sentence, even if the unit is derived from a proper name.

8. Plurals are used as required when writing unit names. For example, henries is plural for henry. The following exceptions are noted:

Singular	Plural
lux	lux
hertz	hertz
siemens	siemens

With these exceptions, unit names form their plurals in the usual manner.

9. No hyphen or space should be left between a prefix and the unit name. In three cases the final vowel in the prefix is omitted: megohm, kilohm, and hectare.

10. The symbol should be used in preference to the unit name because unit symbols are standardized. An exception to this is made when a number is written in words preceding the unit, for example, we would write *ten meters,* not *ten m.* The same is true the other way, for example, 10 m, not 10 meters.

4.6.2
Multiplication and Division

1. When writing unit names as a product, always use a space (preferred) or a hyphen.

Correct usage

newton meter or newton-meter

2. When expressing a quotient using unit names, always use the word *per* and not a solidus (/). The solidus, or slash mark, is reserved for use with symbols.

Correct	Not correct
meter per second	meter/second

3. When writing a unit name that requires a power, use a modifier, such as squared or cubed, after the unit name. For area or volume, the modifier can be placed before the unit name.

Correct	Not correct
millimeter squared	square millimeter

4. When expressing products using unit symbols, the center dot is preferred.

Correct

N·m for newton meter

5. When denoting a quotient by unit symbols, any of the following methods are accepted form:

Correct

m/s or m·s^{-1} or $\dfrac{m}{s}$

In more complicated cases, negative powers or parentheses should be considered. Use m/s^2 or m·s^{-2} but not m/s/s for acceleration; use kg·m^2/(s^3·A) or kg·m^2·s^{-3}·A^{-1} but not kg·m^2/s^3/A for electric potential.

4.6.3
Numbers

1. To denote a decimal point, use a period on the line. When expressing numbers less than 1, a zero should be written before the decimal marker.

Example

15.6
0.93

2. Since a comma is used in many countries to denote a decimal point, its use is to be avoided in grouping data. When it is desired to avoid this confusion, recommended practice calls for separating the digits into groups of three, counting from the decimal to the left or right, and using a small space to separate the groups.

Correct and recommended procedure

6.513 824   76 851   7 434   0.187 62

4.6.4
Calculating with SI Units

Before we look at some suggested procedures that will simplify calculations in SI, the following positive characteristics of the system should be reviewed.

Only one unit is used to represent each physical quantity, such as the meter for length, the second for time, and so on. The SI metric units are *coherent;* that is, each new derived unit is a product or quotient of the fundamental and supplementary units without any numerical factors. Since coherency is a strength of the SI system, it would be worthwhile to demonstrate this characteristic by using two examples. Consider the use of the newton as the unit of force instead of pound-force (lbf). It is defined by Newton's second law, $F = ma$. It is the force that imparts an acceleration of one meter per second squared to a mass of one kilogram. Thus,

$$1 \text{ N} = (1 \text{ kg})(1 \text{ m/s}^2)$$

Consider also the Joule, a unit that replaces the British Thermal Unit, calorie, foot-pound-force, electronvolt, and horsepower-hour to stand for any form of energy. It is defined as the amount of work done when an applied force of one newton acts through a distance of one meter in the direction of the force. Thus,

$$1 \text{ J} = (1 \text{ N})(1 \text{ m})$$

To maintain the coherency of units, however, time must be expressed in seconds rather than minutes or hours, since the second is the base unit. Once coherency is violated, then a conversion factor must be included and the advantage of the system is diminished.

But there are certain units *outside* SI that are accepted for use in the United States, even though they diminish the system's coherence. These exceptions are listed in Tab. 4.7.

Calculations using SI can be simplified if you

1. Remember that all fundamental relationships such as the following still apply, since they are independent of units.

$$F = ma \quad KE = \frac{1}{2}mv^2 \quad E = RI$$

2. Recognize how to manipulate units and gain a proficiency in doing so. Since Watt = J/s = N·m/s, you should realize that N·m/s = $(N/m^2)(m^3/s)$ = (pressure)(volume flow rate).

3. Understand the advantage of occasionally adjusting all variables to base units. Replacing N with kg·m/s^2, Pa with kg·m^{-1}·s^{-2}, and so on.

4. Develop a proficiency with exponential notation of numbers to be used in conjunction with unit prefixes.

$$1 \text{ mm}^3 = (10^{-3} \text{ m})^3 = 10^{-9} \text{ m}^3$$

$$1 \text{ ns}^{-1} = (10^{-9} \text{ s})^{-1} = 10^9 \text{ s}^{-1}$$

4.7

Special Characteristics

A term that should be avoided when using SI is "weight." Frequently we hear statements such as "The man weighs 100 kg." A better statement would be "The man has a mass of 100 kg." To clarify any confusion, let's look at some basic definitions.

Table 4.7 Non-SI units accepted for use in the United States

Quantity	Name	Symbol	SI equivalent
Time	minute	min	60 s
	hour	h	3 600 s
	day	d	86 400 s
Plane angle	degree	°	π/180 rad
	minute	′	π/10 800 rad
	second	″	π/648 000 rad
Volume	liter	L*	10^{-3} m^3
Mass	metric ton	t	10^3 kg
	unified atomic mass unit	u	1.660 57 × 10^{-27} kg (approx)
Land area	hectare	ha	10^4 m^2
Energy	electronvolt	eV	1.602 × 10^{-19} J (approx)

*Both "L" and "l" are acceptable international symbols for liter. The uppercase letter is recommended for use in the United States because the lowercase "l" can be confused with the numeral 1.

First, the term *mass* should be used to indicate only a quantity of matter. Mass, as we know, is measured in kilograms against an international standard.

Force, as defined previously, is measured in newtons. It denotes an acceleration of one meter per second squared to a mass of one kilogram.

The acceleration of gravity varies at different points on the surface of the earth as well as distance from the earth's surface. The accepted standard value of gravitational acceleration is 9.806 650 m/s^2.

Gravity is instrumental in measuring mass with a balance or scale. If you use a beam balance to compare an unknown quantity against a standard mass, the effect of gravity on the two masses cancels out. If you use a spring scale, mass is measured indirectly, since the instrument responds to the local force of gravity. Such a scale can be calibrated in mass units and be reasonably accurate when used where the variation in the acceleration of gravity is not significant.

In the English gravitational system, the unit "pound" is sometimes used to denote both mass and force. We will use the convention that pound-mass (lbm) is a unit of mass and pound-force (lbf) is a unit of force. Thus, pound-mass can be directly converted to the SI unit kilogram, and pound-force units convert to newtons. Another English unit describing mass is the slug.

A word of caution when using English gravitational units in Newton's second law ($F = ma$). The combination of units, lbf, lbm, and ft/s^2, is not a coherent (consistent) set that is, 1 lbf imparts an acceleration of 32.174 ft/s^2 to 1 lbm rather than 1 ft/s^2 required for coherency. A coherent set of English units is lbf, slug, and ft/s^2 because 1 lbf = 1 slug $\times$ 1ft/s^2. You can convert mass quantities from lbm to slugs before substituting into $F = ma$. The slug is 32.174 times the size of the pound-mass (1 slug = 32.174 lbm).

The following example problem clarifies the confusion that exists in the use of the term *weight* to mean either force or mass. In everyday use, the term *weight* nearly always means mass; thus, when a person's weight is discussed, the quantity referred to is mass.

Example problem 4.1   A "weight" of 100.0.kg (the unit itself indicates mass) is suspended by a rope (see Fig. 4.3). Calculate the tension in the rope in newtons when the mass is lifted vertically at constant velocity and the local gravitational acceleration is (a) 9.807 m/s^2 and (b) 1.63 m/s^2 (approximate value for the surface of the moon).

Theory   Tension in the rope when the mass is at rest or moving at constant velocity is

$F = mg$

where g is the local acceleration of gravity and m is the mass of object.

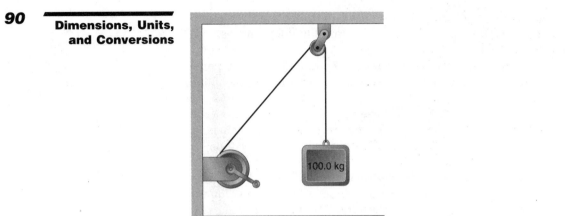

Figure 4.3

Assumption   Neglect the mass of the rope.

Solution

(a) For $g = 9.807$ m/s^2 (given to four significant figures)

$$F = (100.0 \text{ kg})(9.807 \text{ m/s}^2)$$

$$= \underline{0.980\ 7 \text{ kN}}$$

(b) For $g = 1.63$ m/s^2

$$F = (100.0 \text{ kg})(1.63 \text{ m/s}^2)$$

$$= \underline{0.163 \text{ kN}}$$

4.8

Conversion of Units   Although the SI system is the international standard, there are many other systems in use today. It would be fair to say that most of the current work force of graduate engineers has been schooled using terminology such as slugs, pound-mass, pound-force, and so on, and a very high percentage of the total United States population is more familiar with degrees Fahrenheit than degrees Celsius.

For this reason and because it will be some time before the SI system becomes the single standard in this country, you must be able to convert between unit systems.

Four typical systems of mechanical units presently being used in the United States are listed in Tab. 4.8. The table does not provide a complete list of all possible quantities; it is presented to demonstrate the different terminology that is associated with each unique system. If a physical quantity is expressed in any system, it is a simple matter to convert the units from that system to another. To do this, the basic unit conversion must be known and a logical unit analysis must be followed.

Table 4.8 Mechanical units

Quantity	Absolute system		Gravitational system	
	MKS	CGS	Type I	Type II
Length	m	cm	ft	ft
Mass	kg	g	slug	lbm
Time	s	s	s	s
Force	N	dyne	lbf	lbf
Velocity	$m \cdot s^{-1}$	$cm \cdot s^{-1}$	$ft \cdot s^{-1}$	$ft \cdot s^{-1}$
Acceleration	$m \cdot s^{-2}$	$cm \cdot s^{-2}$	$ft \cdot s^{-2}$	$ft \cdot s^{-2}$
Torque	$N \cdot m$	$dyne \cdot cm$	$lbf \cdot ft$	$lbf \cdot ft$
Moment of inertia	$kg \cdot m^2$	$g \cdot cm^2$	$slug \cdot ft^2$	$lbm \cdot ft^2$
Pressure	$N \cdot m^{-2}$	$dyne \cdot cm^{-2}$	$lbf \cdot ft^{-2}$	$lbf \cdot ft^{-2}$
Energy	J	erg	$ft \cdot lbf$	$ft \cdot lbf$
Power	W	$erg \cdot s^{-1}$	$ft \cdot lbf \cdot s^{-1}$	$ft \cdot lbf \cdot s^{-1}$
Momentum	$kg \cdot m \cdot s^{-1}$	$g \cdot cm \cdot s^{-1}$	$slug \cdot ft \cdot s^{-1}$	$lbm \cdot ft \cdot s^{-1}$
Impulse	$N \cdot s$	$dyne \cdot s$	$lbf \cdot s$	$lbf \cdot s$

Mistakes can be minimized if you remember that a conversion factor simply relates the same physical quantity in two different unit systems. For example, 1.0 in and 25.4 mm each describe the same length quantity. Thus, when using the conversion factor 25.4 mm / in to convert a quantity in inches to millimeters, you are multiplying a factor that is not numerically 1 but is physically one. This fact allows you to readily avoid the most common error, that of using the reciprocal of a conversion. Just imagine that the value in the numerator of the conversion must describe the same physical quantity as that in the denominator. When so doing, you will never use the incorrect factor 0.304 8 ft/m, since 0.304 8 ft is clearly not the same length as 1 m.

Example prob. 4.2 demonstrates a systematic procedure to use when performing a unit conversion. The construction of a series of horizontal and vertical lines separating the individual quantities will aid the thought process and help ensure a correct unit

Figure 4.4
Military systems are designed in accordance with SI standards.
(General Dynamics, Pomona Division.)

analysis. In other words, the units to be eliminated will cancel out, leaving the desired results. The final answer should be checked to make sure it is reasonable. For example, the results of converting from inches to millimeters should be approximately 25 times larger than the original number.

Example problem 4.2   Convert 6.7 in to millimeters.

Solution   Write the identity

$$6.7 \text{ in} = \frac{6.7 \text{ in}}{1}$$

and multiply by conversion factor.

$$\frac{6.7 \text{ in}}{1} \left| \frac{25.4 \text{ mm}}{1 \text{ in}} \right. = 1.7 \times 10^2 \text{ mm}$$

Example problem 4.3   Convert 85.0 lbm/ft^3 to kilograms per cubic meter.

Solution

$$85.0 \text{ lbm/ft}^3 = \frac{85.0 \text{ lbm}}{1 \text{ ft}^3} \left| \frac{1^3 \text{ ft}^3}{(0.304 \; 8)^3 \text{ m}^3} \right| \frac{0.453 \; 6 \text{ kg}}{1 \text{ lbm}}$$

$$= 1.36 \times 10^3 \text{ kg/m}^3$$

Example problem 4.4   Determine the gravitational force (in newtons) on an auto with a mass of 3 645 lbm. The acceleration of gravity is known to be 32.2 ft/s^2.

Solution A   Force, mass, and acceleration of gravity are related by $F = mg$.

$$m = \frac{3645 \text{ lbm}}{1} \left| \frac{1 \text{ kg}}{2.204 \; 6 \text{ lbm}} \right. = 1 \; 653.36 \text{ kg}$$

$$g = \frac{32.2 \text{ ft}}{1 \text{ s}^2} \left| \frac{0.304 \; 8 \text{ m}}{1 \text{ ft}} \right. = 9.814 \; 6 \text{ m/s}^2$$

$$F = mg = (1 \; 653.36 \text{ kg})(9.814 \; 6 \text{ m/s}^2) = 16 \; 227 \text{ N} \cong 16.2 \text{ kN}$$

Note: Intermediate values were not rounded to final precision, and we have used either exact or conversion factors with at least one more significant figure than contained in the final answer.

Solution B

$$F = mg = \frac{3 \; 645 \text{ lbm}}{1} \left| \frac{32.2 \text{ ft}}{1 \text{ s}^2} \right| \frac{1 \text{ kg}}{2.204 \; 6 \text{ lbm}} \left| \frac{0.304 \; 8 \text{ m}}{1 \text{ ft}} \right.$$

$$= 16 \; 227 \text{ N} \cong 16.2 \text{ kN}$$

Note: It is often convenient to include conversions with the appropriate engineering relationship in a single calculation.

Example problem 4.5   Convert a mass flow rate of 195 kg/s (typical of the airflow through a turbofan engine) to slugs per minute.

Solution

$$195 \text{ kg/s} = \frac{195 \text{ kg}}{1 \text{ s}} \left| \frac{1 \text{ slug}}{14.594 \text{ kg}} \right| \frac{60 \text{ s}}{1 \text{ min}} = 802 \text{ slug/min}$$

Example problem 4.6   Compute the power output of a 225-hp engine in (a) British thermal units per minute and (b) kilowatts.

Solution

(a) $225 \text{ hp} = \dfrac{225 \text{ hp}}{1} \left| \dfrac{2.546\ 1 \times 10^3 \text{ Btu}}{1 \text{ hp·h}} \right| \dfrac{1 \text{ h}}{60 \text{ min}}$

$\qquad = 9.55 \times 10^3 \text{ Btu/min}$

(b) $225 \text{ hp} = \dfrac{225 \text{ hp}}{1} \left| \dfrac{0.745\ 70 \text{ kW}}{1 \text{ hp}} \right. = 168 \text{ kW}$

The problem of unit conversion becomes more complex if an equation has a constant with hidden dimensions. It is necessary to work through the equation converting the constant K_1 to a new constant K_2 consistent with the equation units.

Consider the following example problem given with English units.

Example problem 4.7   The velocity of sound in air (c) can be expressed as a function of temperature (T):

$$c = 49.02\sqrt{T}$$

where c is in feet per second and T is in degrees Rankine.
   Find an equivalent relationship when c is in meters per second and T is in kelvins.

Procedure

1. First, the given equation must have consistent units; that is, it must have the same units on both sides. Squaring both sides we see that

$$c^2 \text{ft}^2/\text{s}^2 = 49.02^2 T^\circ R$$

It is obvious that the constant $(49.02)^2$ must have units in order to maintain unit consistency. (The constant must have the same units as c^2/T.)
   Solving for the constant,

$$(49.02)^2 = c^2 \frac{\text{ft}^2}{\text{s}^2} \left[\frac{1}{T^\circ R} \right]$$

$$= \frac{c^2}{T} \left[\frac{\text{ft}^2}{\text{s}^2{}^\circ R} \right]$$

2. The next step is to convert the constant 49.02² ft²/(s²°R) to a new constant that will allow us to calculate c in meters per second given T in kelvins. We recognize that the new constant must have units of square meters per second squared-kelvin.

$$\frac{(49.02)^2 \text{ ft}^2}{1 \text{ s}^2{}^{\circ}R} = \frac{(49.02)^2 \text{ ft}^2}{1 \text{ s}^2{}^{\circ}R} \left| \frac{(0.304\ 8)^2 \text{ m}^2}{1 \text{ ft}^2} \right| \frac{9{}^{\circ}R}{5\ K} = \frac{401.84 \text{ m}^2}{1 \text{ s}^2 K}$$

3. Substitute this new constant 401.84 back into the original equation

$$c^2 = 401.84T$$

$$c = 20.05\sqrt{T}$$

where c is in meters per second and T is in kelvins.

4.9

Key Terms and Concepts

Physical quantities	Fundamental dimensions
Units	Derived dimensions
Base units	Absolute system
Supplemental units	Gravitational system
Derived units	Metric system
Dimensions	Symbols

Problems

4.1 Using the correct number of significant figures, convert the following physical quantities to the proper SI units.
- (*a*) 645 lbm
- (*b*) 98.2 °F
- (*c*) 4.75 × 10² acres
- (*d*) 55 × 10² gal
- (*e*) 110.0 × 10³ gal/h
- (*f*) 88 ft/s
- (*g*) 285 hp
- (*h*) 2025 in
- (*i*) 1.255 × 10² ft³/min

4.2 Convert the following to SI units. Use correct significant figures.
- (*a*) 750.5 Btu/min
- (*b*) 65.2 hp·h
- (*c*) 4.500 × 10³ mi
- (*d*) 1.00 × 10² mi/h
- (*e*) 225 lbf
- (*f*) 8.255 × 10⁴ lbm/ft³
- (*g*) 1.955 atm
- (*h*) 212°F
- (*i*) 5280.0 ft

4.3 Convert as indicated giving answer with proper significant figures.
- (*a*) 85.5 in to centimeters
- (*b*) 505 L to cubic feet
- (*c*) 78.8°C to degrees Fahrenheit
- (*d*) 10 750 bushels to cubic centimeters
- (*e*) 65.5 × 10⁵ Btu/h to kilowatts

4.4 Convert as indicated giving answer with proper sign figures.
- (*a*) 7.550 × 10³ km to feet
- (*b*) 285 K to degrees Fahrenheit
- (*c*) 6.85 × 10⁴ ft lbf to joules
- (*d*) 14.7 lbf/in² to pascals
- (*e*) 77.7 slug/ft³ to grams per cubic centimeter

4.5 Using the rules for expressing SI units, express each of the following in correct form if given incorrectly.
- (*a*) 11.5 cm's
- (*b*) 475 N
- (*c*) 9.5 m/s/s
- (*d*) 5000 K
- (*e*) 8,000 pa
- (*f*) 25 amp
- (*g*) 100.1 m m
- (*h*) 62.5 j
- (*i*) 300 degrees kelvin

4.6 Using the rules for expressing SI units, express each of the following in correct form if given incorrectly.

  (a) 53 m per sec   (d) 25 farads   (g) .5 mm
  (b) 101C   (e) 1 000 N   (h) 40 nM
  (c) 75 A's   (f) 48 Kg   (i) 8050 N/m/m

4.7 What force in newtons would be required to lift with uniform velocity a 275 lbm vise under the following conditions?

  (a) Acceleration of gravity is 32.2 ft/s^2
  (b) Acceleration of gravity is 9.80 m/s^2
  (c) Acceleration of gravity is 25 ft/s^2

4.8 Determine the acceleration of gravity (meters per second squared) if the force required to lift at uniform velocity a 4.000×10^2 kg object is

  (a) 1295 lbf
  (b) 1055 lbf
  (c) 585 N

4.9 What work is done to lift a 25.0×10^2 lbm object 4.50×10^1 ft vertically if the acceleration of gravity is 9.807 m/s^2? Express answer in joules. *Note:* Work = force $\times$ distance traveled in force direction

4.10 Determine the engine power (kilowatts) and horsepower required to move an automobile on level ground if the resistance of the tires and air resistance is 35.0 lbf. The auto is traveling at 88.0 ft/s. *Note:* Power = (force)(velocity).

4.11 The density of water at 70°F is about 1.936 slug/ft^3. First determine the volume of a 50.00 ft diameter spherical tank. Then compute the mass of water contained if the tank is full. Express volume in cubic meters and mass in kilograms.

4.12 A cylindrical tank 8.0 ft long and 6.0 ft in diameter with hemispherical ends is placed so its longitudinal axis is horizontal. How many gallons of gasoline are in the tank if the fluid level is 4.5 ft above the bottom of the tank? What is the mass of the gasoline in kilograms if its specific gravity is 0.67?

4.13 Determine the total force in the ground exerted by the air above a sports field 55.0 by 100 yd if the air pressure measured at field level is 2088 lbf/ft^2. Express your answer in newtons.

4.14 Compute the mass (in kilograms) of gravel stored in a rectangular feeder bin 17.5 by 25.0 ft. The depth of the gravel is 15.0 ft and its density is 97 lbm/ft^3. If the feeder bin is elevated on supports 20.0 ft above the ground, what vertical load in newtons does the gravel place on the supports? Assume standard gravitational attraction.

4.15 Shelled corn is often piled on the ground because of insufficient storage facilities. Compute the diameter of a pile in feet, number of bushels in the pile, volume of the pile in cubic meters, and the mass of the pile in kilograms if the height of the pile is 17.5 ft. *Hint:* The angle of repose for dry shelled corn is about 12 degrees and its density is 56 lbm/bushel.

4.16 Use a spreadsheet to do the computation in Prob. 4.15 for a range of heights from 10.0 to 30 ft in increments of 0.50 ft. Prepare a printed copy so that a grain dealer can estimate the amount of corn in a pile by measuring its height.

4.17 Do Prob. 4.16 but modify the spreadsheet so that the dealer can simply measure the circumference of the base of the pile instead of the height. Use a range of circumferences from 35.0 to 275 ft in increments of 5.00 ft.

4.18 The Darcy-Weisbach friction formula for pipes allows us to compute the frictional energy loss per unit mass (ft lbf/lbm or J/kg) of a fluid flowing through a pipe by

$$h_L = (f)\left(\frac{L}{D}\right)\left(\frac{v^2}{2g}\right)$$

where h_L = energy loss per unit mass

f = friction factor, dimensionless

L = length of pipe

D = diameter of pipe

v = average velocity of fluid

g = acceleration of gravity

Calculate h_L in J/kg and in ft·lbf/lbm for water flowing at 25.0 ft/s through a 1.00-in ID cast-iron pipe, 275 ft long. Assume $f = 0.040$ for this situation.

4.19 A weir is used to measure flow-rates in open channels. For a rectangular weir, the expression can be written

$$Q = 5.35\ LH^{3/2}$$

where

Q = discharge rate, ft³/s

L = length of weir, ft

H = height of fluid above crest, ft

(a) Determine a new constant so the expression can be applied with Q in gallons per hour, and L and H in inches.

(b) Write a computer program or prepare a spreadsheet to produce a table of values of Q in both ft³/s and gallons per hour, with L (inches) and H (inches). Use values of L that range from 1 to 15 ft in 1 inch increments. Let H range from 6 to 36 in increments of 2 inches.

4.20 For certain conditions, the law of conservation of energy can be written

$$V = 4.429\sqrt{h}$$

where

V = velocity, m/s

h = distance, m

Determine a new constant so that the equation is valid for h in ft and V in miles per hour.

4.21 The specific fuel consumption of an engine may be written as

$$sfc = \frac{2545}{\eta Q}$$

where

sfc = specific fuel consumption, lbm/hp·h

η = thermal efficiency, dimensionless

Q = heat of combustion per unit mass, Btu/lbm

Determine a new constant for Q in J/kg and sfc in kg/kW·h.

4.22 The universal law of gravity for the force of attraction between two masses may be written as

$$F = (6.673 \times 10^{11}) \frac{m_1 m_2}{r^2}$$

where F is the force of attraction, N

m_1, m_2 are masses, kg
r is separation distance, m

Determine a new constant for F in lbf, m_1 and m_2 in lbm, and r in mi.

4.23 A hollow aluminum sphere, 260 mm in diameter (outside) with a wall thickness of 3.10 mm.
   (a) Compute the outside surface area (square millimeters).
   (b) Compute the mass of the ball (kilograms).
   (c) Determine whether the ball will float in water (density of 62.4 lbm/ft^3).

4.24 A piece of oak (density = 47 lbm/ft^3) has been cut into the shape of a right rectangular pyramid with a base of 4.00 in by 3.00 in and a height of 8.00 in. Determine
   (a) The outside surface area (square inches).
   (b) The mass in lbm.
   (c) The height of the base above the surface of water (density of 62.4 lbm/ft^3), assuming the object will float point down.

4.25 An object in the shape of a right cone with a base diameter of 290 mm and a height of 320 mm is floated point down in alcohol (density of 49 lbm/ft^3). The base of the cone extends 75 mm above the surface of the alcohol.
   (a) What is the density of the cone in kg/m^3?
   (b) What is the mass of the cone in kg?
   (c) What is the surface area in square millimeters of the cone that is below the surface of the alcohol (wetted area)?

Index